Student Solutions

Mathematics for Elementary School Teachers

Ricardo D. Fierro
California State University, San Marcos

Prepared by

Ricardo D. Fierro
California State University, San Marcos

Scott Fallstrom
Mira Costa College, Oceanside

BROOKS/COLE
CENGAGE Learning

Australia • Brazil • Japan • Korea • Mexico • Singapore • Spain • United Kingdom • United States

© 2013 Brooks/Cole, Cengage Learning

ALL RIGHTS RESERVED. No part of this work covered by the copyright herein may be reproduced, transmitted, stored, or used in any form or by any means graphic, electronic, or mechanical, including but not limited to photocopying, recording, scanning, digitizing, taping, Web distribution, information networks, or information storage and retrieval systems, except as permitted under Section 107 or 108 of the 1976 United States Copyright Act, without the prior written permission of the publisher.

For product information and technology assistance, contact us at
**Cengage Learning Customer & Sales Support,
1-800-354-9706**

For permission to use material from this text or product, submit all requests online at **www.cengage.com/permissions**
Further permissions questions can be emailed to
permissionrequest@cengage.com

ISBN-13: 978-1-133-36374-3
ISBN-10: 1-133-36374-1

Brooks/Cole
20 Davis Drive
Belmont, CA 94002-3098
USA

Cengage Learning is a leading provider of customized learning solutions with office locations around the globe, including Singapore, the United Kingdom, Australia, Mexico, Brazil, and Japan. Locate your local office at: **www.cengage.com/global**

Cengage Learning products are represented in Canada by Nelson Education, Ltd.

To learn more about Brooks/Cole, visit
www.cengage.com/brookscole

Purchase any of our products at your local college store or at our preferred online store
www.cengagebrain.com

Printed in the United States of America
1 2 3 4 5   18 17 16 15 14

Table of Contents for the Solutions

Chapter 1 Problem Solving and Reasoning   1

Chapter 2 Sets, Numeration, and Addition and Subtraction with Whole Numbers   13

Chapter 3 Multiplication and Division with Whole Numbers   23

Chapter 4 Elementary Number Theory and Integers   31

Chapter 5 Rational Numbers and Fractions   39

Chapter 6 Decimals, Real Numbers, and Percents   49

Chapter 7 Algebra and Functions   57

Chapter 8 Descriptive Statistics   65

Chapter 9 Probability   73

Chapter 10 Introduction to Geometry   83

Chapter 11 Measurement   89

Chapter 12 Triangles and Quadrilaterals   99

Chapter 13 Coordinate Geometry and Plane Transformations   111

Acknowledgements.

From Ricardo D. Fierro: I would like to express my deepest gratitude to Scott Fallstrom for the many hours he devoted to collaborating on the solution manual. It was a pleasure to work with him.

From Scott Fallstrom:

To Manesseh: Amazingly, when I woke up the morning after this was finished, you were still there. I wish to thank Marissa and Haley for putting up with all of the long days, weeks, and months (and sometimes crankiness) that it took to get this finished. Thank you to the Dew Crew for tolerating my absence at so many events. To my parents for always pushing me to succeed. And to Dr. Fierro and the Cengage crew for the opportunity.

Chapter 1 – Section 1.1

1. Arithmetic sequence with initial term 1 and common difference 3:
   a. Words: Beginning with the number 1, the next term is obtained by adding 3 to the previous term.
   b. Table (just list a few terms):
n	1	2	3	4
y	1	4	7	10
   c. Algebra: $y = 3n - 2$, where y is the nth term.
   d. Graph:

   Graph of the Arithmetic Sequence 1, 4, 7, 10, ...

3.
   c. The sequence in table form looks like:
n	1	2	3	4
y	–5	2	9	16

5.
   f. Pick two points on the line, then calculate the slope: (1, 4) and (3, 10) lie on the line. The slope is:
      $$\text{slope} = \frac{\text{rise}}{\text{run}} = \frac{10-4}{2-1} = \frac{6}{2} = 3.$$
   g. The common difference of the arithmetic sequence equals the slope of the line through the points on the graph of the arithmetic sequence.

7.
   a. This sequence starts with 8 students at 1 table. Then, as one more table is added, 2 more students are able to sit. So the sequence representing the number of students is: 8, 10, 12, 14, ... This is arithmetic with first term of 8 and common difference of 2. An equation to represent this would be $y = 8 + 2(n-1) = 8 + 2n - 2 = 2n + 6$, where n is the number of tables joined together and y is the number of students able to be seated. So when 37 tables are joined together, $n = 37$, and we can find y:
   $y = 2 \cdot 37 + 6 = 74 + 6 = 80$. So when 37 tables are joined together, 80 students can be seated.
   b. If we know the number of students is 185, that's the y-value. So $y = 185$, and we can find n: $185 = 2n + 6$, $179 = 2n$, $n = 179 \div 2 = 89.5$. Since we can't have half of a table, we'll need 90 tables to seat the 185 students.

8. Rico started with $11. After that, he deposited a fixed amount, d dollars, each month into the account and ended with $195. Using the arithmetic sequence equation because the amount is increasing each month by a fixed amount, we get: $195 = 11 + d(n-1)$, so $184 = d(n-1)$
   $184 = d(n-1)$. Since n must be a whole number, we could just plug in some values for n and find different possible values for d. If $n = 2$, then $d = 184$. If $n = 3$, then $2d = 184 \Rightarrow d = 92$. If $n = 4$, then $3d = 184 \Rightarrow d = 61\frac{1}{3}$ ... which doesn't make sense since we can't invest exactly $\frac{1}{3}$ of a dollar. If $n = 5$, then $4d = 184 \Rightarrow d = 46$. Continuing in this way other values could be found but each should be checked to see if it is a realistic amount of money.

11. Solve simpler problems, look for a pattern. The dots represent players, and the line segments represent games played between the players.

    | 2 players | 3 players | 4 players |
    | 1 game | 1 + 2 games | 1 + 2 + 3 games |

    The pattern suggests if there are 23 players in the tournament, then $1 + 2 + 3 + \ldots + 21 + 22$ games would be played. Using Gauss's method, we would get
    $S = 1 + 2 + 3 + 4 + \ldots + 21 + 22$
    $S = 22 + 21 + \ldots + 1$
    Adding the two equations gives us:
    $2S = 22 \times 23$
    $2S = 506$
    $S = 506 \div 2$
    $S = 253$
    So 253 games would be played.

13. $37^2 = 1369$;
    $337^2 = 113{,}569$;
    $3337^2 = 1{,}1135{,}569$;
    $33{,}337^2 = 1{,}111{,}355{,}569$.
    So it seems that each new '3' in the base will create one additional 1 to the left, and one additional 5

between the 3 and the 69. Using this pattern,
$3,333,337^2 = 11,111,135,555,569$.

15.
- b. The conjecture is "An odd number can be written as the sum of two consecutive whole numbers."

17.
- a. 34, 35, 36 has 3 numbers in the list and 36 − 34 = 2. 34, 35, 36, 37 has 4 numbers in the list and 37 − 34 = 3. 34, 35, 36, 37, 38 has 5 numbers in the list and 38 − 34 = 4. So we would subtract, then add 1.
- b. Using part (a), the number of terms in 34, 35, 36, …, 621 is 621 − 34 + 1 = 588.

19. The right-hand side of the equations form a geometric sequence 64, 16, 4, 1, … with common ratio 1/4.
$4^3 = 64$
$4^2 = 16$
$4^1 = 4$
$4^0 = 1$

20. The right-hand side of the equations form an arithmetic sequence with common difference − 4.
$3 \times 4 = 12$
$2 \times 4 = 8$
$1 \times 4 = 4$
$0 \times 4 = 0$
$-1 \times 4 = -4$
$-2 \times 4 = -8$

23.
- a. Start skip-counting with 20's until the next term would exceed the divisor (85). 20, 40, 60, 80; and 85 − 80 = 5. So $85 \div 20 = 4 \text{ R} 5$.
- b. Start skip-counting with 15's until the next term would exceed the divisor (52). 15, 30, 45; and 52 − 45 = 7. So $52 \div 15 = 3 \text{ R} 7$.
- c. Start skip-counting with 12's until the next term would exceed the divisor (64). 12, 24, 36, 48, 60; and 64 − 60 = 4. So $64 \div 12 = 5 \text{ R} 4$.

25. These equations seem to be adding consecutive odd numbers and getting perfect squares. The next perfect square is 25, and 16 + 9 = 25. So $1+3+5+7+9 = 5^2$, and the next one would be $1+3+5+7+9+11 = 6^2$. It seems the sum of the first n odd numbers is n^2.

28.
- a. Yes, the sequence could be arithmetic with a common difference of 1.
- b. No, because the ratio is not common (or constant). The first ratio is 2 and the second is 1.5.

31.
- d. The terms seem to add the two previous terms to get to the next one. So the next three likely terms are 34, 55, and 89.
- e. This sequence seems to be combined from 2 different sequences: 11, 13, 15, … and 4, 6, 8, 10, … . It seems someone put these two together to create a new sequence and if that's the case, then the next three terms would be 17, 12, and 19.

34. b.

n	1	2	3	4
y	−3	1	5	9

- e. To find how many terms are less than 513, we can use the previous equation. Any term corresponding to position n is given by $y = 4n - 7$, which we want to be less than 513.
$y < 513 \Rightarrow 4n - 7 < 513 \Rightarrow 4n < 520$
$\Rightarrow n < 130$. So term number 130 has the value of 513, meaning there are 129 terms less than 513.

36. Marissa could be correct if the sequence is progressing in a way that adds one to the coefficient on a, while also multiplying the constant terms by 2, 3, 4, 5. Using this reasoning, the term after would be 600 + 6a.

39.
- a. In order to use Gauss's method, we need to find the number of terms. Since the sequence is increasing by 2 each time, then the last term could be written with the arithmetic sequence equation as: $410 = 204 + 2(n-1)$. Solving this gives $n = 104$ and there are 104 terms in the sequence. The sum will be
$S = \dfrac{(204 + 410) \times 104}{2} = 31,928$.
- b. In a similar way, $496 = 88 + 8(n-1)$ is solved to obtain $n = 52$. The sum will be
$S = \dfrac{(88 + 496) \times 52}{2} = 15,184$.

40.
- a. 1, 6, 11, 16, … $y = 5n + 1$
0, 5, 10, 15, … $y = 5n$
Both sequences have common difference 5.
The coefficient of n is 5.

b. 5, 8, 11, 14, … $y = 3n + 2$
    3, 6, 9, 12, … $y = 3n$
    Both sequences have common difference 3.
    The coefficient of n is 3.

43. Answers vary.
    For example, $y = 3906.25 \times 2^{n-1}$.
    Here's how we found our answer.
    We need to find a and r such that
    $62,500 = a \times r^{5-1}$. Then $62,500 = a \times r^4$. There are many choices for a and r. Choose $r = 2$. Then $62,500 = a \times 2^4$. Then $62,500 = a \times 16$. Then $a = 62,500/16 = 3906.25$.
    So one possible geometric sequence is described by the equation $y = 3906.25 \times 2^{n-1}$.

47. a. Table of values:

Year	Value (exact $)	Value (to the penny)
0	25,000	25,000
1	21,250	21,250
2	18,062.5	18,062.5
3	15,353.125	15,353.13
4	13,050.15625	13,050.16
5	11,092.6328125	11,092.63

    a. The largest depreciation in value is in the first year where it drops $3750. All the rest are smaller as they take a percentage of a smaller value.
    b. The ratio of any two consecutive values is 0.85; this shows that the sequence is geometric with common ratio of 0.85.

49. It may help to look at the pattern here, as it shows the first shape needs 4 toothpicks. To create the next shape, remove the bottom toothpick, and slide up two toothpick squares (formed with 2 toothpicks on top, 2 toothpicks on bottom, and 3 vertical toothpicks). The new number of toothpicks is $4 - 1 + 7 = 10$. The next shape would remove the bottom 2 and slide up 3 toothpick squares (formed with 3 on top, 3 on bottom, and 4 vertical). The new number is $10 - 2 + 10 = 18$. This continues and our sequence of toothpicks is: 4, 10, 18, 28, … . It's like an arithmetic sequence but we're adding 6, then 8, then 10, etc. So the differences form an arithmetic sequence and it may help to re-write the terms with a table:

Term	Number of Toothpicks	Number of Toothpicks Written out Differently
1	4	4
2	10	4 + 6
3	18	4 + 6 + 8
4	28	4 + 6 + 8 + 10
5	40	4 + 6 + 8 + 10 + 12
6	54	4 + 6 + 8 + 10 + 12 + 14

    The 6th shape can be seen from the table, but to find any shape we could just write the terms out and use Gauss's method as well.

52.
    a. The sequence marked as "A" is the arithmetic sequence.
    b. You can draw a straight line through the points.

Section 1.2

2. Let's organize the results in a table, then look for a pattern:

# cuts	1	2	3
# pieces	2	3	4

    The pattern suggests a stick cut into 3094 pieces requires 3093 cuts ($3094 - 1 = 3093$).

5. Making a table is a good way to start here and knowing that we could have pennies (1¢), nickels (5¢), or dimes (10¢) to use is helpful. Start with a table showing the 21¢ with all pennies and then re-combine with different methods from there. The table shows there are exactly 9 ways to make 21¢ with coins.

Ways	Pennies	Nickels	Dimes
1	21	0	0
2	16	1	0
3	11	2	0
4	11	0	1
5	6	3	0
6	6	1	1
7	1	4	0
8	1	2	1
9	1	0	1

6. The pages of a book form an arithmetic sequence since they increase by one. The common difference is 1 and the initial term is 143, so we could write an equation: $y = 143 + 1(n - 1)$, $y = n + 142$. Choose 851 for y and solve: $851 = n + 142$, $n = 709$. There are 709 pages in the chapter.

9. If you are only making one call, then you'll have $9.41 for that one phone call. There is always a $4 service fee, leaving $5.41 for phone calls. At 25¢ per minute, the number of minutes would be $0.25n \leq 5.41 \Rightarrow n \leq 21.64$. So you'd only have enough money for 21 minutes of long distance calls (typically calls are rounded up to the next whole minute).

11. $51 - 3 = 48$, so 48 cm of wood were used to make the cube. A cube has 12 edges. Each edge has

length n cm. So $12n = 48$. Then $n = 4$. So the length of each edge of the cube is 4 cm.

13. This type of problem can be reasoned through by thinking about the worst case scenario. There are 6 possible prizes, say A, B, C, D, E, F. The mother could purchase 6 cereal boxes and get all 6 different prizes. Then she could purchase 6 more boxes and get A, B, C, D, E, and F again. So purchasing 12 prizes does not guarantee three identical prizes for her triplets. But the prize in the 13th box of cereal must match one of the pairs, giving her three identical prizes. Thus, she must purchase 13 boxes of cereal to guarantee obtaining three identical prizes.

17. The product of two consecutive numbers is 214,832. Let's use the Guess and Check strategy.
$400 \times 401 = 160,400$ too low
$450 \times 451 = 202,950$ too low
$465 \times 466 = 216,690$ too high
$463 \times 464 = 214,832$ yes
The pages are 463 and 464.

20.
   a. The hour hand on a cuckoo clock completes one revolution in 12 hours. This is a rate of 360° per 12 hours, or 30° per hour. Since 1 hour = 60 minutes, this could be written as 30° per 60 minutes, or 0.5° per minute.
   b. The minute hand turns at rate of one revolution per hour, which would be 360° per hour. This is 360° per 60 minutes, which equals 6° per minute.

21. It may be easiest to think of this as blocks of 5 DVDs: the 4 you pay for and the 1 free one. So 5 DVDs would cost $12 since one was free. Take the 46 DVDs he rented and divide into groups of 5: $46 \div 5 = 9\ R1$. There were 9 blocks at $12 per block, and then 1 more single at $3. Total cost would be: $9 \times \$12 + 1 \times \$3 = \$111$.

23.
   a.
   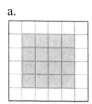
   b. We will make a table and look for a pattern.

nth shape	1	2	3	4
y, # of white tiles	8	12	16	20

$y = 8 + 4(n - 1)$, or $y = 4n + 4$, where y is the number of white tiles needed for the nth shape.

   c. The number of dark tiles needed for the nth shape is n^2. We need to find the smallest value of n such that $n^2 > 3(4n + 4)$, or $n^2 > 12n + 12$.

n	n^2	$12n + 12$
10	100	132
11	121	144
12	144	156
13	169	168

Then 169 dark tiles are needed.

28.
   a. It seems Polly subtracted $851 - 143$ to get 708.
   b. Solve a simpler problem such as a chapter begins on page 2 and ends on page 4. How many pages are in the chapter? It has pages 2, 3, 4, so it has 3 pages. We quickly see that subtracting $(4 - 2 = 2)$ gives the incorrect answer. Polly needs to add one more to the difference. $708 + 1 = 709$.

32.
   a. In the piece of wood that is 62 feet long, he can cut 20 three-foot logs (since $20 \times 3 = 60$).
   b. Since each of the 20 logs requires one cut, it will take him $20 \times 4 = 80$ minutes. However, if there were 63 feet of log to cut, then he could cut 21 three-foot logs in the same amount of time because the last cut would make 2 three-foot logs.

35. Write an equation. Let d equal the number of raffle tickets sold for the Disneyland trip. Then $43 - d$ raffle tickets were sold for the golf clubs. Then
$2d + 5(43 - d) = 161$
$2d + 215 - 5d = 161$
$3d = 54$
$d = 18$.
18 raffle tickets to Disneyland were sold.

37. XOOO = 49¢ and XXXOO = 91¢, so
XOOO
XOOO = 147¢ $(3 \times 49 = 147)$
XOOO
Substitute three X and two O for 91¢:

      O
91¢ + OOO = 147¢. Then
      OOO
O
OOO = 56¢. Then O = 8¢.
OOO
Then X + 8¢ + 8¢ + 8¢ = 49¢.
So X = 25¢ and O = 8¢.

42.
  a. With four draws, you could get a spade, heart, diamond, and club in the worst possible case. For this reason, 4 or fewer draws do not guarantee drawing two cards with the same suit. But the fifth card must match one of the suits already drawn. Therefore, drawing 5 cards would guarantee two cards with the same suit.
  b. With 8 draws, you could get two spades, two hearts, two diamonds, and two clubs in the worst possible case. For this reason, 8 or fewer draws do not guarantee drawing three cards with the same suit. But the ninth card must match one of the pairs of suits already drawn. Therefore, drawing 9 cards would guarantee three cards with the same suit.

45. The perimeter of the fence is at most 34 ft.

w	l	perimeter	area
5	10	$2 \times 5 + 2 \times 10 = 30$	50
5	11	$2 \times 5 + 2 \times 11 = 32$	55
5	12	$2 \times 5 + 2 \times 12 = 34$	60
6	11	$2 \times 6 + 2 \times 11 = 34$	66

    The fence should have dimensions $6' \times 11'$.

47. Iggy is faster and has to travel 110 cm at 8 cm per hour, then it would take Iggy $110 \div 8 = 13.75$ hours. Zed is slower and needs to travels just $110 - 25 = 85$ cm to the finish line due to the head start. $85 \div 6 = 14\frac{1}{6}$, so it takes Zed $14\frac{1}{6}$ hours to reach the finish line. Iggy wins the race!

Section 1.3

3. The expression $5n - 3$ represents the number of coins Tanya has.

4.
  a. A diagram would look like this:

  b. Let c represent the number of pencils Cory has. Then Veronica has $3c + 5$ pencils.

6. Answers vary.
  a. Ken has 3 more stamps than Lana. Write a variable expression to represent the relationship, where Lana has n stamps.
  b. Ken has 5 fewer stamps than Lana. Write a variable expression to represent the relationship, where Lana has n stamps.
  c. Ken has 4 more than 3 times as many stamps as Lana. Write a variable expression to represent the relationship, where Lana has n stamps.
  d. Ken has 3 fewer than 4 times as many stamps as Lana. Write a variable expression to represent the relationship, where Lana has n stamps.

11.
  a. Answers vary. $3 + 5 = 8$; $5 + 7 = 12$; $33 + 59 = 92$. It seems that every time we add 2 odd numbers together, we get an even number. The conjecture would be "The sum of two odd numbers is an even number."
  b. An even number has the form $2 \times$ (whole number), or $2k$ for some whole number k. An odd number has the form $2 \times$ (whole number) $+ 1$, or $2k + 1$ for some whole number k. Let a and b represent any two odd numbers. Then $a = 2m + 1$ and $b = 2n + 1$ for some whole numbers m and n.

  Then $a + b = 2m + 1 + 2n + 1$
  $= 2m + 2n + 2$
  $= 2(m + n + 1)$

  $m + n + 1$ is a whole number, so $a + b$ equals 2 times a whole number. So $a + b$ is an even number.

13.
  a. The table would look like:

c	10	11	12	13
m	2	3	4	5

  b. Mike has 8 fewer, so Mike would have $c - 8$.
  c. An equation that would appropriately model the situation is $m = c - 8$.

16. The original radius is r cm, the new radius is $R = 2r$. Then $\pi R^2 = \pi (2r)^2 = \pi (4r^2) = 4(\pi r^2)$, so the new area is 4 times as large as the original area of the circle.

18. To have "2 more than n" means we need to have n objects in the first place. So "2 more than n" is appropriately represented by $n + 2$ rather than $2 + n$.

20.
  a. Brian has 7 times as many pencils as Sandra.
  e. Brian has 3 fewer than 5 times as many pencils as Sandra.

21.
  a. $48 - 5 = 43$. Melanie has $43 more than Steve.

22.
   a. $23 \div 7 = 3$ R2. Kate collected 2 more than 3 times as many butterflies as Ellen did.

25.
   a. The directions would be something like: "Pick a number. Add 5 to it. Multiply the result by 9. Add 3. Subtract the original number. Divide the result by 4. Tell me your final result."
   b. If you algebra, we can simplify this result: $[9(n+5)+3-n] \div 4 = 2n+12$. So the final result will be $2n + 12$, and in order to get n we just subtract 12 and then divide by 2. Or equivalently, we could divide by 2 and then subtract 6.

27.
   a. The n tells you how many socks Sam has.
   b. The $3n - 4$ tells you how many socks Max has.

31. After 5 months, Tom has $5d + 200$ dollars in his savings account, where d is the fixed amount he deposited each month. The check he receive in the mail was worth $3(5d + 200)$ dollars. When he deposited the check, he had $5d + 200 + 3(5d + 200) = 3000 + 400$ dollars.
So $5d + 200 + 15d + 600 = 3400$
$20d + 800 = 3400$
$20d = 2600$
$d = 130$
He deposited $130 each month for 5 months.

32. Answers vary. Ken and Montika have 73 pairs of shoes together. Montika has 3 fewer than threee times as many as Ken. How many pairs of shoes does each one have?

33.
   a. Let j represent Jerry's age and a represent Amanda's age. Then $a = j + 3$.

36. Let n represent the smallest addend in the sum. Then we can calculate:
$n + (n + 1) + (n + 2) \ldots + (n + 27) = 686$
$28n + (1 + 2 + 3 + \ldots + 27) = 686$
$28n + (27 \times 28)/2 = 686$
[because $1 + 2 + 3 + \ldots + 27 = (27 \times 28)/2$].
$28n + 378 = 686$
$28n = 308$
$n = 11$
The smallest addend in the sum is 11.

39. Let n represent the number of child tickets sold. Then $2.25n$ represents the number of adult tickets sold. So $n + 2.25n = 390$
$3.25n = 390$
$n = 390 \div 3.25$
$n = 120$, the number of child tickets sold.
$2.25n = 2.25(120) = 270$, the number of adult tickets sold. $8(270) + 3(120) = \$2520$, the revenue due to ticket sales that day.

40.
   a. $4^2 - 3^2 = 16 - 9 = 7 = 4 + 3$;
$5^2 - 4^2 = 25 - 16 = 9 = 5 + 4$;
$6^2 - 5^2 = 36 - 25 = 11 = 6 + 5$.
   b. It seems that a good conjecture would be "The difference of two consecutive squares is the sum of the two numbers that were squared."
   c. If our first number is n, then our next consecutive number would be $n + 1$. Squaring and subtracting would result
$(n + 1)^2 - n^2 = n^2 + 2n + 1 - n^2$
$= 2n + 1$
$= n + (n + 1)$
So $(n + 1)^2 - n^2$ is the sum of the two consecutive numbers $n + 1$ and n.
   d. The difference is $2n + 1$, so we could solve for n if we are given the difference.
$2n + 1 = 853$
$2n = 852$
$n = 426$. So the smaller number is 426 and the larger number is 427. Checking this we see that $427^2 - 426^2 = 182,329 - 181,476 = 853 = 427 + 426$.

45.
   a. The next 3 terms would be:
$2 \cdot 5 - 5 = 5$
$2 \cdot 6 - 6 = 6$
$2 \cdot 7 - 7 = 7$
   b. The algebraic notation for this would be $2n - n = n$.
   c. Yes, algebra makes the pattern obvious.

48. You may ask the student if putting in different values for a will make the equation true. Is it true when $a = 3$? Is it true when $a = 5$? Showing how different numerical values could change the equation from true to false can help show that a is a variable.

49.
   a. $336,400 - 149,505 = 186,895$, so the memoir of early military campaigns sold for $186,895 more than an early draft of Napoleon's will.
   b. $336,400 \div 149,505 = 2$ R$37,390$, so the memoir of early military campaigns sold for $37,390 more than twice the price of an early draft of Napoleon's will.

52.
  a. Answers vary. 3 × 5 = 15, 7 × 3 = 21; 5 × 9 = 45. The examples support the conjecture "the product of two odd numbers is an odd number."
  b. An odd number can be written in the form $2a + 1$ for some whole number a. Let m and n represent any two odd numbers. Then $m = 2x + 1$ for some whole number x and $n = 2y + 1$ for some whole number y. Then
  $mn = (2x + 1)(2y + 1)$
  $= 2x2y + 2x + 2y + 1$
  $= 2(2xy + x + y) + 1$
  $2xy + x + y$ is a whole number, so $mn = 2q + 1$, where q is the whole number $2xy + x + y$.
  So mn is an odd number.

Section 1.4

1.
  a. Since all dogs chase cats, then the dog oval (A) will be inside the "things that chase cats" oval (B), which is diagram III.
  b. Some motorists do not drive fast, if A is the set of people who drive fast and B is the set of motorists, then diagram IV shows our statement.
  c. No politicians tell the truth is best reflected in diagram I, since this shows separate ovals for politicians (A) and people who tell the truth (B).
  d. If you play baseball, then you are an athlete. So baseball player oval (A) is inside the athlete oval (B) meaning diagram III is the best choice.

2. There are many correct pictures here depending on how your ovals are labeled.
  a. Some dogs chase cats.

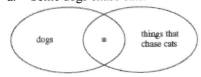

  b. Some students do not drink coffee.

  c. There is at least one politician who tells the truth.

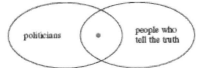

  d. If you are a vegetarian, then you like spinach.

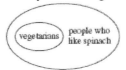

4. Lisa's reasoning is not correct. Mike may know how to multiply, but he doesn't have to be a mathematician. Since the conclusion does not automatically follow, the argument is invalid.

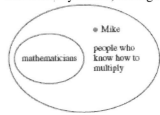

5. An Euler diagram is shown below, illustrating that the argument is valid since the conclusion follows automatically from the premises.

7. Let n be an odd number. Then $n = 2k + 1$ for some whole number k. Then $n + 4 = 2k + 1 + 4 = 2(k + 2) + 1$. Since k is a whole number, then $k + 2$ is a whole number, making the quantity $2(k + 2)$ an even number. Then $2(k + 2) + 1$ is an odd number. This shows that $n + 4$ can be expressed as 2 times a whole number plus 1, so $n + 4$ is an odd number.

9. If 5 divides n, then this means $n = 5k$ for some whole number k. Multiplying both sides of this equation by 6 will give the equation $6n = 30k$. But $30k$ can be written as $10(3k)$. This means $6n = 10(3k)$, and since $3k$ is a whole number, it means that 10 divides $6n$.

11. This argument is invalid. There is no guarantee that "You" have to be inside the ring that represents people who win races.

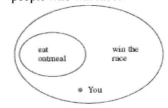

14.
  a. This is the fallacy of the inverse
  b. This is the fallacy of the converse.
  c. This is the fallacy of the converse.
  d. This is the fallacy of the inverse.

17. Answers vary.
  a. $P \to Q$ is true and P is true. Then make sure your statement Q is true.
     An example of a statement Q is "3 is a prime number."
  b. $P \to Q$ is false and P is true. Then make sure your statement Q is false. An example of such a statement Q is "3 is not a prime number."

19. The first Venn diagram seems to support the conclusion, but the second Venn diagram does not support the conclusion. Therefore, the conclusion does not inescapably follow since it's possible to satisfy the premises but fail the conclusion. So Taylor's reasoning is incorrect.

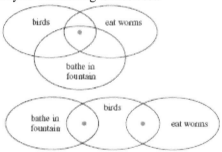

21.
  a. Answers vary. We just need one rectangle that is not a square. One is shown below (there are many others).

  b. We need one whole number that less than or equal to 10; the number 5 is a counterexample (there are many others).
  c. We need to find whole numbers a, b, and c such that $a^2 + b^2 = c^2$. Since $3^2 + 4^2 = 5^2$, we have a counterexample (many other counterexamples exist).
  d. We need to find a solution to $x^2 - 5x + 6 = 0$. There are only two counterexamples here: 2 and 3 since $x^2 - 5x + 6 = (x-2)(x-3)$.
  e. We need to find a factor of 24 that is not an even number. There are two counterexamples: 1 and 3.

22. For a conditional statement to be false, the hypothesis needs to be true and the conclusion false.
  a. If we let $n = 2$, then $5 \times n = 5 \times 2 = 10$ which is not odd.
  b. If we let $a = 2$ and $b = 1$, then $(2-1)^2 = 2^2 - 1^2$; this implies $1 = 3$ which is false.
  c. If we let $x = 500$, then $2x + 6 = 1006$, which is not less than 1000.
  d. If we let $a = 400$ and $b = 2$, then $a > b$, but $ab < 500$ is false since $ab = 800$.
  e. If we let $x = 500$, then x is greater than 5 but is not an odd number.

24. Each statement has the form "If P, then Q." The statement is true in the case P is true and Q is true, or in the case P is false. The statement is false if P is true and Q is false.
  a. This is true, since it is of the form $P \to Q$ and both P and Q are true.
  b. This is true, since it is of the form $P \to Q$ and P is false. The truth value of Q is irrelevant.

28.
  a. The passengers did not board the train and they found a taxi.
  b. He did not send an email or he did not purchase stamps at the post office.
  c. Mark does have a job or he does not walk to school.
  d. Mary did not call her mother and Jim did ride the bike.

32.
  a. There are some computers in the library.
  b. The lights in the office never flicker.
  c. Some dogs do not chase cats.
  d. Some rats play football.
  e. There are some fleas on the cat.
  f. No whole numbers are negative integers.
  g. Some cars do not require fuel.

33. Answers vary.
  a. Choose statements P and Q that are both true or both false. For example, P could be $3 + 5 = 8$ and Q could be $7 \times 4 = 28$.
  b. Choose P that is false and Q that is true. For example, P could be $3 + 5 \neq 8$ and Q could be $7 \times 4 = 28$.

35. The Venn diagram shows that the conclusion does not inescapably follow. Therefore, the argument is invalid.

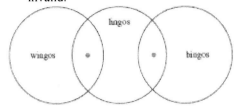

38. Suppose m is an odd number and n is an even number. Then $m = 2a + 1$ and $n = 2b$ for some whole numbers a and b. Then $m + n = 2a + 1 + 2b = 2(a + b) + 1$. We know $a + b$ is a whole number. This means $m + n$ can be expressed in the form 2 times a whole number plus 1. So $m + n$ is an odd number.

39. This argument is invalid; it is the fallacy of the inverse.

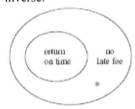

41. This argument is valid; it follows the Law of Detachment.

44. This argument is invalid. We can represent it with an Euler diagram. Notice how the diagram shows the premises to be true (1) All mathematicians use calculators and (2) Some teachers are not mathematicians. Yet in this picture, some teachers use calculators, so the conclusion doesn't follow automatically from the premises. Therefore, the argument is invalid.

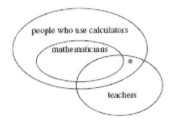

Chapter 1 Review

2. $6 \times 3 = 18$; $8 \times 5 = 40$; $12 \times 7 = 84$. A conjecture would be " the product of an even number and an odd number is an even number."

4.
  a. The pattern repeats after 4 terms, and $345 \div 4 = 86$ R1. This means that the 345th term would have the same value as the 1st term, which is "W".

6.
  c. The number of dark bricks in an n-by-n patio can be found many ways.

n	3	4	5	6
# of dark bricks	8	12	16	20

  The second row of the table contains multiples of 4. Then the number of dark bricks in an n-by-n patio is given by the expression $4(n - 1)$.

8.
  a. 15, 30, 45, 60. $74 - 60 = 14$. So $74 \div 15 = 4$ R14.

12.
  c. The number of corners seems to increase by 2 each time, so the formula for the number of corners on the perimeter of the nth shape would be $y = 4 + 2(n - 1)$, or $y = 2n + 2$.

14. If 3307 is the 15th term, and 3 is the initial term, then an equation representing this would be:
$3307 = 3 + d(15 - 1)$
$3304 = d(14)$
$236 = d$
So the common difference would be 236.

18.
  a. The common difference is 4.

21. Use the Guess and Check strategy to find two numbers n and $n + 1$ such that $n(n + 1)$ equals 46,010. The pages are 214 and 215.

25. Harry decorates 5 cookies per minute and Larry does 3 per minute. So let L represent the number of minutes Larry decorates, so then L – 5 is the number of minutes Harry worked. Then the number of cookies Larry decorated would be 3L and the number of cookies Harry decorated would be 5(L – 5). Together, they decorated 191 cookies, so the equation is:
$3L + 5(L - 5) = 191$
$8L = 216$
$L = 27$.
This means Larry worked for 27 minutes and Harry worked for 22 minutes. So Larry decorated $3 \cdot 27 = 81$ cookies and Harry decorated $5 \times 22 = 110$ cookies.

28. Think of these as blocks of 7 DVDs (6 paid for and 1 free). $121 \div 7 = 17$ R2. So he paid for the last 2, as well as 6 DVDs in each of the 17 blocks for a total of $17 \times 6 + 2 = 104$ DVD rentals at $4 each. The total cost would be $416.

31. The diagram is shown below:

    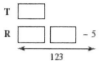

    $123 + 5 = 128$, $128 \div 2 = 64$. Tony has 64 marbles.

34. No, the definition of the variable is not acceptable. Saying that n = Cindy implies that n represents a person, not a number. The student probably meant that n = the number of coins that Cindy has.

36.
    b. $48 \div 14 = 3$ R6, so Mitchell has 6 more than 3 times as many coins as Joey.

41.
    a. The statement is <u>true</u> because it is an 'or' statement, and only one of the statements must be true in an 'or' statement.
    b. This statement is <u>true</u> because both parts are true and in an 'and' statement, both parts must be true.

42.
    a. Some birds don't eat worms.

45. A possible Venn diagram is shown. The conclusion "Jerry plays soccer" does not inescapably follow, so the argument is invalid.

    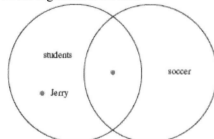

51.
    a. $P \to Q$ is false and Q is false. Any statement P that is true would suffice. An example of such a statement P is "3 is a prime number."
    b. $P \to Q$ is true and P is true. Any statement Q that is true would suffice. An example of such a statement Q would be "$2 + 3 = 5$".

55. The argument is valid by the Law of Detachment.

Chapter 1 Test

2. $3 + 4 + 5 = 12$; $6 + 7 + 8 = 21$; $10 + 11 + 12 = 33$. It seems that every sum of 3 consecutive whole numbers is divisible by 3. The conjecture is: "The sum of 3 consecutive whole numbers is divisible by 3."

4.
   a. This is arithmetic, so the equation would be: $y = 1 + 3(n - 1)$ or $y = 3n - 2$
   b. To see if 567 is a term in the sequence, use the equation from (a): $567 = 3n - 2$, $569 = 3n$, $n = 189\frac{2}{3}$. No, 567 is not a term in this sequence.
   c.   d.

   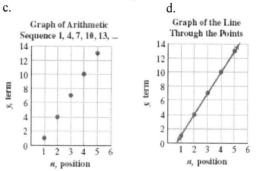

   e. The slope of the line is 3 (rise/run = 6/2 = 3).
   f. The slope of the line is the common difference of the sequence.

8. Since this pattern repeats every 4 terms, we can use the remainder to find which of the terms will correspond to the 105th term. $105 \div 4 = 26$ R1, so the 105th term will be the same as the 1st term, which is H.

11. Notice that the first table seats 10 people, and adding another table increase that number to 14. Continuing to add tables would continue adding 4 to each term so this is an arithmetic sequence with an equation of: $y = 10 + 4(n - 1)$, or $y = 4n + 6$. We need to seat 73 people, so use the equation
    $73 = 4n + 6$
    $67 = 4n$
    $16.75 = n$
    Because we can't have a fraction of a table, we will need 17 tables to seat the 73 people.

12. Define d to be the amount of time (in minutes) that Don spends painting, so Marc spends $d + 20$ minutes painting. Since Marc can paint 4 square feet of wall per minute, he paints $4(d + 20)$ square feet of all and Don paints 6 square feet per minute for $6d$ square feet of wall. The equation formed is: $6d + 4(d + 20) = 720$. Solving for d, we get $d = 64$. So Don spends 64 minutes painting the room.

14.
    a. Algebra is the use of variables to represent unknown quantities or relationships, as well as manipulating the variables.
    b. A variable is a letter or symbol most often representing a number or numerical quantity.
    c. An equation is a statement that sets two expressions equal to each other.

16.
    a. Since it's close to 3 times the number of child tickets, we can start with $3n$. But since they needed to sell five fewer adult tickets to get here, then this amount should be increased by 5 to get the number of adult tickets to $3n + 5$.
    b. Solving this problem with algebra requires an equation:
$$n + (3n + 5) = 253$$
$$4n + 5 = 253$$
$$4n = 248$$
$$n = 62$$
    c. So there were 62 child tickets sold. From part (b) above, if 62 child tickets were sold, then there were $3 \cdot 62 + 5 = 191$ adult tickets sold. Each adult ticket was $11 and each child ticket was $6, so the total revenue was:
$$11 \cdot 191 + 6 \cdot 62 = \$2473.$$

17.
    a. Mitchell has 29 more coins than Joey.
    b. $41 \div 12 = 3 \text{ R} 5$, so Mitchell has 5 more than 3 times as many coins as Joey.

21. The Euler diagram follows:

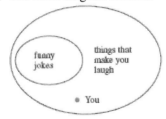

In this diagram, the dot indicates that you laughed. Since this could be outside the funny jokes oval, the argument is invalid. (This is the fallacy of the converse.)

22. The Euler diagram follows:

In this diagram, the dot indicates that you do not eat granola each day. Since this must be outside the people who are athletes oval, the argument is valid. (This is reasoning with the Law of Contraposition.)

Chapter 2 – Section 2.1

2.
  b. $\sim B = \{3, 4, 5\}$
  c. $A \cap B = \varnothing$

3. The Venn diagram for this problem is:

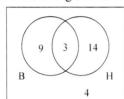

  a. 4 students did not like either sport.
  b. $9 + 14 = 23$ students liked exactly one sport.

5.
  a. The set of students who like history and math.
  b. The set of students who like only math and history
  c. The set of students who like only math.
  d. The set of students who like only chemistry or only history.

7. The Venn diagram represents the survey of 32 customers. The Venn diagram suggests there were $11 + 5 + 20 = 36$ people surveyed, which exceeds the number of people actually surveyed.

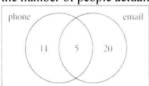

There's certainly an error here so either the customers responded inaccurately or the advertiser made a mistake in counting the responses.

9.
  d. The conjecture would be "$A \cup B = B \cup A$ for any sets A and B."

10.
  d. The conjecture would be "$A \cap B = B \cap A$ for any sets A and B."

13. No, it is not possible. The Venn diagram suggests there would be 20 students enrolled in history or math rather than 25 students in both combined as stated in the problem.

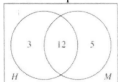

15. A Venn diagram can help organize the given information and make it easier to visualize the relationships:

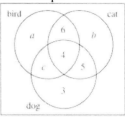

We need to determine the sum $c + 4$.
8 people said they own exactly one pet,
so we know $a + b + 3 = 8$. Then $a + b = 5$.
There are 27 people who own a bird, cat, or dog, so
$a + b + c + 6 + 4 + 5 + 3 = 27$
$5 + c + 6 + 4 + 5 + 3 = 27$
$c = 4$
Then $c + 4 = 4 + 4 = 8$. So 8 people owned a bird and dog.

18. The Venn diagram would be:

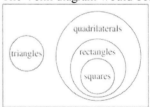

19. Yes, this is possible and we just need one set to have one element and the other to have 5 elements. For example, $A = \{n\}$ and $B = \{1, 2, 3, 4, 5\}$.

21.
  b. $n(A \cap B) = 0$ means $A \cap B = \varnothing$, so A and B are disjoint (they do not have any elements in common).
  d. $n(\sim A) = 0$ means $\sim A = \varnothing$, which implies there are no elements that are not in A. This would mean that every element in the universe belongs to A. In other words, $A = U$ (A is the universe).

22. The symbol $\cup$ looks like the first letter of the word "union."

25.
  a. $A \cup B = \{1, 2, 3, 4, 6, 7\}$; $A \cap B = \{3, 4\}$; $A - B = \{6\}$; $\sim A = \{1, 2, 5, 7, 8\}$
  b. $A \cup B = \{1, 2, 3, 4, 5, 6\}$; $A \cap B = \varnothing$; $A - B = \{2, 4, 6\}$; $\sim A = \{1, 3, 5, 7, 8\}$

27.
  a. True, since 7 is an element of {7, 8, 9}.
  b. True, since every element in {a, b} is in {a, b, c}, and the element c is not in {a, b}.
  c. True, since 5 is not an element of {2, 3, 4}.
  d. True, since every element in {1, 2, 5} is in {1, 2, 5}.
  e. False, since the two are equal, one can't be a proper subset of the other.
  f. True, since the number 3 is not an element of { {1}, {2}, {3} }.

29.
  a. The cardinality is 4; the elements are 1, 6, 0, and the infinite set {7, 8, 9, ...}.
  b. The cardinality is 1; the element is $\varnothing$.
  c. The cardinality is 1; the element is { {4,5} }.

31.
  b. $n(A \cap B) \neq 0$ means $A \cap B \neq \varnothing$, which implies A and B have at least one element in common.
  d. $n(\sim A) \neq 0$ means $\sim A \neq \varnothing$, which implies there is at least one element in the universe that is not in A. We would know that A is not the universal set. Symbolically, $A \neq U$.

33. The Venn diagram is:

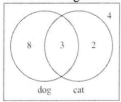

  a. There are 4 students who do not have a dog or a cat, shown as the number 4 outside the circles.
  b. There are 10 students who have only a dog or only a cat.

36. Let C represent the set of students who read *Charlotte's Web* and let T represent the set of students who read *Trumpet of the Swan*. Then $n(C) = 15$ and $n(T) = 20$. The number of students who read *Charlotte's Web* or *Trumpet of the Swan* must be less than or equal to the total number of students in the class, so $n(C \cup T) \leq 23$.
We know
$n(C \cup T) = n(C) + n(T) - n(C \cap T)$.
$n(C) + n(T) - n(C \cap T) \leq 23$.
$15 + 20 - n(C \cap T) \leq 23$.
$35 - n(C \cap T) \leq 23$.
$12 \leq n(C \cap T)$.
At the very least, 12 students have read both books.

38.
  a. A, since $A - B = \sim B \cap A$
  b. $\sim B$ or $\varnothing$, since $A \cap B = A - \sim B$ and $A \cap B = A - \varnothing$

40. Yes. Notice that there is a one-to-one correspondence between {2, 4, 6, 8, ...} and the counting numbers {1, 2, 3, 4, ...} and as well as a one-to-one correspondence between {1, 2, 3, 4, ...} and {5, 10, 15, 20, ...}. So a in A corresponds to $a \div 2$ in the set of counting numbers, which in turn corresponds to the number $(a \div 2) \cdot 5$ in B. This means the element a of A corresponds to the element $\frac{5}{2}a$ of B. Then $2 \leftrightarrow 5$, $4 \leftrightarrow 10$, and $6 \leftrightarrow 15$.

41.
  a. $B \cup \varnothing = B$
  b. $B - \varnothing = B$
  c. This might remind you of subtraction since $n - 0 = n$ or addition since $n + 0 = n$.

46. The minimums and maximums are based on the possible intersections. Venn diagrams can help.
  a. 22. The minimum number of students occurs when one set is a subset of the other. In this case, the minimum number of students is 22 – which occurs when all "Jeopardy" watchers also saw "Wheel of Fortune".

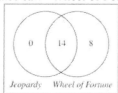

  b. 36. The maximum number of students occurs when the two sets are disjoint. In this case, the maximum number of students is 36 – which occurs when all "Jeopardy" watchers didn't see "Wheel of Fortune".

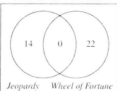

Section 2.2

3.
  b. The number line shows 286 (the dot) is closer to 300 than to 200, so we would round to 300.

4.
    b. 145,241,083

6. Mayan numerals
    a.

7. Babylonian numerals
    a.

8. Since these are base five numerals, any time we have five or more of an object, we need to regroup
    a. 2001_{five}. We need to regroup the 6 units as 1 long and 1 unit, making 5 longs. These 5 longs can be regrouped as 1 flat making 5 flats. The 5 flats can be regrouped as 1 cube making 2 cubes.

10. Using base three blocks:
    a.

15. It may help to solve some simpler problems. $7108 \times 10^2 = 7108 \times 100 = 710{,}800$ which has 6 digits. $7108 \times 10^3 = 7108 \times 1000 = 7{,}108{,}000$, which has 7 digits. So it seems that you take the 4 digits to start with and add the exponent of 10 to find the total number of digits. In this case, there would be $4 + 23 = 27$ digits.

17.
48,000 rounded to the nearest *one* equals 48,000. 48,002 rounded to the nearest *ten* equals 48,000. 48,031 rounded to the nearest *hundred* equals 48,000. 48,287 rounded to the nearest *thousand* equals 48,000. We could have rounded to the nearest one, ten, hundred, or thousand.

19.
    a. Answers vary. An example of a value *n* for this would need to be a number that rounds to 8,000, since 4,258 rounds to 4,000; 7980 = *n* would work.

20.
    a. There are 7 groups of 20 and 13 groups of 1; this is the base ten numeral $7 \times 20 + 13 \times 1 = 140 + 13 = 153$.

    b. There is 1 group of 360, 2 groups of 20 and 3 groups of 1; this is the base ten numeral $1 \cdot 360 + 2 \cdot 20 + 3 \cdot 1 = 403$.

21.
    a. $54 = 2 \times 20 + 14$.

    b. Since the number is less than 360, we can start with groups of 20. $348 \div 20 = 17 \text{ R}8$. Then $348 = 17 \times 20 + 8$.

22.
    a. This numeral in base ten would be $11 \cdot 3600 + 0 \cdot 60 + 1 \cdot 1 = 39{,}601$.
    b. This numeral in base ten would be $30 \cdot 3600 + 1 \cdot 60 + 12 \cdot 1 = 108{,}072$.

23.
    a. 54 is smaller than 60, so this is just 54 groups of 1. The Babylonian numeral is:

    b. $348 \div 60 = 5 \text{ R}48$, so this is 5 groups of 60 and 48 groups of 1. The Babylonian numeral is:

27.
    a. $134_{\text{five}} = 1 \cdot 100_{\text{five}} + 3 \cdot 10_{\text{five}} + 4 \cdot 1_{\text{five}}$ or $134_{\text{five}} = 1 \cdot 10^2_{\text{five}} + 3 \cdot 10^1_{\text{five}} + 4 \cdot 10^0_{\text{five}}$.

29. b. $450B_{\text{twelve}} = 4 \cdot 12^3 + 5 \cdot 12^2 + B \cdot 12^0 = 7643$

35. The answer "the underlined digits have different values" seems to imply a deeper understanding of place value concepts because values take into account both place value of the digit and the digit itself.

37. Both the Mayan and base ten numeration systems are place value systems, although the Mayan is a modified base system since the same factor is not used at each step. Further, the Mayan is a vertical system (read top to bottom) while the base ten system is horizontal (read left to right). The Mayan system uses fewer symbols, but because of the regrouping actually has more digits (need a unique representation of all values from 0 to 19 instead of just 0 to 9).

40.
  a. 328,052,000,211

43.
  b. 0

45.
  a. They each represent 2 groups of something
  b. They have different values
  c. The commas break a numeral into periods to make it easier to read.

48.
  a. 45 thousand, 36
  b. forty-five thousand, thirty-six
  c. $40,000 + 5,000 + 30 + 0 + 6$
  d. $4 \cdot 10,000 + 5 \cdot 1000 + 0 \cdot 100 + 3 \cdot 10 + 6 \cdot 1$
  e. $4 \cdot 10^4 + 5 \cdot 10^3 + 0 \cdot 10^2 + 3 \cdot 10^1 + 6 \cdot 10^0$

52. Answers vary.
  a. 3, 15, 6 or 2, 25, 6
  b. 623, 50, 1, or 623, 49, 11

55.
  b. 425 to the nearest hundred is 400.

57.
  a. $842 \div 15 \approx 56.13$. So 842 is between 56×15 and 57×15. $0.13 < 0.5$ and $56 \times 15 = 840$, so 842 rounded to the nearest multiple of 15 is 840.

59.
  a. 10 groups of 3600 and 11 groups of 1 give a base ten numeral of:
  $10 \times 3600 + 0 \times 60 + 11 \times 1 = 36,011$

60.
  a. $600_{\text{nine}} + 70_{\text{nine}} + 8_{\text{nine}}$

61.
  a. $678 \div 5^4 = 1 \text{ R} 53$; $53 \div 5^2 = 2 \text{ R} 3$. So the base five representation of the numeral is 10203_{five}.

Section 2.3

1.
  a. We need to create two disjoint sets, one with 5 objects and the other with 2 objects. Then we join the sets and count how many total objects.
  ● ● ● ● ●   ● ● → ● ● ● ● ● ● ●
  Counting shows there are 7 objects, so $5 + 2 = 7$.

2.
  a. We need to create 6 objects and then show 2 of them being taken away so we can count what is left. There are 4 objects left, so $6 - 2 = 4$.

3.
  a. The number line model shows that we end up at 8, so $2 + 6 = 8$.

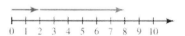

4. Answers vary, but it should involve breaking up or decomposing to form easy combinations.
  a. $21 + 74 = 95$

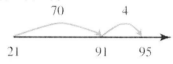

11.
  a. $k - 15 = 22$
  $k - 15 + 15 = 22 + 15$
  $k = 22 + 15$
  $k = 37$

13.
  a. $12 = n + 3$. Paul has 3 more coins than Maria.

16.
  a. Using the definition of subtraction:
  $n - 18 = 25$
  $n = 25 + 18$
  $n = 43$
  b. addition and subtraction are inverse operations:
  $n - 18 = 25$
  $n - 18 + 18 = 25$
  $n = 25 + 18$
  $n = 43$
  c. Using a fact family.
  $n - 18 = 25 \quad n - 25 = 18$
  $n = 18 + 25 \quad n = 25 + 18$
  The last two equations both lead to $n = 43$.

19. The two related equations that follow from the definition of subtraction are $n = 13 + 4$ and $n = 4 + 13$.

22. Two sentences would be: "John has 6 fewer trading cards than Bob" and "Bob has 6 more trading cards than John."

25. Answers vary. There are 3 ways discussed in this chapter: (i) fact families, (ii) definition of subtraction, and (iii) addition and subtraction are inverse operations.

31.
   a. Using the number line model, we end at 11, so 4 + 7 = 11.

33.
   a. To model comparison with 4 < 6, create two groups of objects, one with 4 objects and another with 6 objects. Then match up the objects in a one-to-one correspondence as much as possible. Whichever group has some leftover objects is the larger number.
   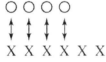
   This shows that 4 < 6.

34. The definition of less than requires us to find a counting number. We know that 5 + 4 = 9, and 4 is a counting number, so by definition of less than, 5 < 9.

35. For number line comparisons, the number to the left is less than the number to the right.
   a. This number line shows 2 < 6.

40.
   a. The appropriate equation is 3 + n = 21; the first equation would have a meaning of "Paul started with some coins and Maria gave him 3."

42.
   a. The number line shows he must drive 3 miles to get home since 8 − 5 = 3.
   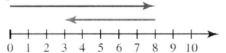

43. Answers vary.
   a. Maria has 8 coins. Frankie has 7 coins. How many coins do they have altogether?

47.
   c. 4 + 6 = 10; 6 + 4 = 10; 10 − 6 = 4; 10 − 4 = 6.

51. The fact family is 57 − m = 29; ⟨57 − 29 = m;⟩ m + 29 = 57; 29 + m = 57.

54. The student added 15 and 4 because 15 + 4 = 19. He started with the 15 and then counted up.

62.
   b. The set {0, 1} is not closed under addition because even though 0 + 0 = 0 which works, 0 + 1 = 1 = 1 + 0 which works, 1 + 1 = 2 and 2 ∉ {0, 1}.
   c. The set {7, 14, 21, 28, …} is closed under addition. Each element in the set can be written as 7k, where k is some counting number. If you take any two of these (7k and 7m) and add them together, you'll get 7k + 7m = 7(k + m), which is another element of the set.

Section 2.4

1.
   a. Step 1, build the minuend 34.
   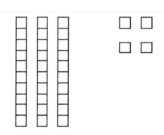

   Step 2, regroup 1 ten as 10 ones.
   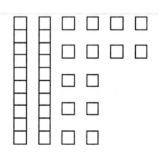

   Step 3, take-away 8 ones
   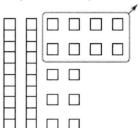

   Then 34 − 8 = 26.

c. Step 1. Build each addend.
Step 2. There are 11 ones. We need to regroup the 11 ones as 1 ten and 1 one.

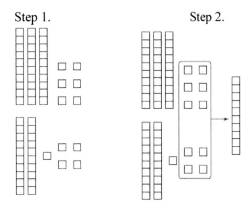

No more regrouping is possible, so 36 + 25 = 61.

3. We have 3 ones and we need to take away 9 ones. There are not enough ones, so we regroup one of the tens as 10 ones. Instead of having 6 tens, we now have 5 tens.

5. Answers vary. Together, the addends here must have 13 ones, 8 tens and 5 hundreds. There are many possibilities such as 537 + 56.

7. Answers vary. There are many possibilities such as 197 + 265.

9.
   a. Since we want no regroupings, then the largest column sum is 9 for each column. This means a must be 5, b must be 3, and c must be 2; so the largest possible addend abc is 532.
   b. If we want one regrouping only and the largest possible value, then regrouping in the hundreds would be best. The largest possible addend abc is 932.
   c. If we want two regrouping only and the largest possible value, then regrouping in the hundreds and tens would be best. The largest possible addend abc is 992.

10.
   a. There are 3 possible partial sums coming from the ones, tens and hundreds.

15.
   a. Using partial sums:
   ```
         3  4  5
      +  9  3  8
      ─────────
            1  3
            7  0
      1  2  0  0
            8  3
      1  2  0  0
      ─────────
      1  2  8  3
   ```

   b. Using the lattice method:

   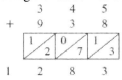

16.
   a. 16 seconds + 45 seconds = 61 seconds, which can be regrouped into 1 minute and 1 second. Then 54 minutes + 12 minutes + 1 minute (from regrouping) = 67 minutes, which can be regrouped into 1 hour and 7 minutes. 3 hours + 4 hours + 1 hour (from regrouping) = 8 hours. A final answer of 8 hours, 7 minutes, 1 second. You could write this like the column method:

	Hours	Minutes	Seconds
	3	54	16
+	4	12	45
	7	66	61
	7	67	1
	8	7	1

17.
   a. Since 1 hour is 60 minutes, we could regroup the 12 hours and 24 minutes into 11 hours and 84 minutes. This makes the rest of the subtraction much easier.

	Hours	Minutes	Seconds
	11	84	25
−	4	56	14
	7	28	11

   The end result is 7 hours, 28 minutes, 11 seconds.

19. Consider the subtraction problem 34 − 18 using the standard subtraction algorithm. 34 = 3 tens, 4 ones would be regrouped as 2 tens, 14 ones in order to have enough ones to take away 8 ones.

24.
   b. First, build each addend. Put the ones together and the tens together. Regroup the 12 tens as 1 hundred and 2 ones.

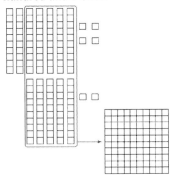

Then there are 1 hundred, 2 tens, and 6 ones. Then 74 + 52 = 126.

   c. Step 1. Build the minuend 48. Then take away 5 ones and 3 tens. There are 1 ten and 3 ones remaining. Then 48 − 35 = 13.

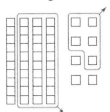

30.
   a. 7 ones + 5 ones = 12 ones. This can be regrouped into 1 ten and 2 ones, so we place the 1 group of ten in the tens place to be added with the other tens later.

35.
   a. Scratch addition:

$$\begin{array}{r} \overset{1}{2}\,\overset{2}{5} \\ \overset{\cancel{6}_0}{}\,\overset{\cancel{7}_2}{} \\ 3\,\overset{\cancel{9}_1}{} \\ +\,4\,\,2 \\ \hline 1\,7\,3 \end{array}$$

36. Answers vary.
   a. 34 − 8 = 36 − 10 = 26

37. Rounded/compatible numbers are not unique, and these are estimates so other estimates are possible with different values.
   a. 43 + 879 ≈ 40 + 880 = 920

40. Partial differences:
   b.
$$\begin{array}{r} 7\,5\,7 \\ -\,8\,3 \\ \hline 4 \\ -\,3\,0 \\ \hline 7\,0\,0 \\ 4 \\ \hline 6\,7\,0 \\ \hline 6\,7\,4 \end{array}$$

41. We'll show the left to right trade-first method

   a.
$$\begin{array}{r} \overset{3}{\cancel{4}}\,\overset{15}{\cancel{5}}\,6 \\ -\,2\,7\,3 \\ \hline 1\,8\,3 \end{array}$$

42.
   c. 131 − 63 = 1 + 130 − 60 − 3
               = 1 + 70 − 3
               = 1 + 67
               = 68

43.
   a. 57 + 43 = 50 + 7 + 43
              = 50 + 50
              = 100

50.
   a.
$$\begin{array}{r} \overset{1\,1\,1}{3\,4\,3}_{\text{five}} \\ +\,3\,2\,4_{\text{five}} \\ \hline 1\,2\,2\,2_{\text{five}} \end{array}$$

   c.
$$\begin{array}{r} \overset{1\,1\,1}{A\,5\,7}_{\text{twelve}} \\ +\,4\,B\,7_{\text{twelve}} \\ \hline 1\,3\,5\,2_{\text{twelve}} \end{array}$$

51.
   a.
$$\begin{array}{r} 3\,\overset{3\,13}{\cancel{4}\,\cancel{3}}_{\text{five}} \\ -\,1\,2\,4_{\text{five}} \\ \hline 2\,1\,4_{\text{five}} \end{array}$$

   c.
$$\begin{array}{r} \overset{3\,15}{4\,\cancel{5}\,A}_{\text{twelve}} \\ -\,B\,7_{\text{twelve}} \\ \hline 3\,6\,3_{\text{twelve}} \end{array}$$

Chapter 2 − Review

1. Answers vary.
   a. $A = \{1, 2, 3\}, 1 \in A$
   b. $A = \{1, 2\}, B = \{1, 2, 3\}, A \subseteq B$

2.
   a. This is true since {7} is an element of the set.
   b. This is false; 8 is an element in the set, but 7 is not an element in the set; {7} is an element in

20

5.
   a. $A = \{a, d, f, h\}$

6.
   b. $n(B) = 10$

9.
   a. The maximum number of elements in the union would occur when the intersection is the smallest possible. So if $X \cap Y = \emptyset$, then the maximum value is $n(X \cup Y) = 6 + 13 = 19$.

11.
   d. $12 + 6 + 6 = 24$ watched exactly one game show.
   e. $4 + 3 + 1 = 8$ watched exactly two game shows.

13. In order the blanks would be: symbols, numerals.

17.
   a. There are no regroupings here as there are not more than 3 of any piece. The number would be represented as the base four numeral 2213_{four}.

18.
   a. $316 = 31$ tens, 6 ones.

21. After 360, all place values multiply by 20. So the next highest place value is $144{,}000 \times 20 = 2{,}880{,}000$.

22.
   a. 71 thousand, 492

24.
   a. 2116

26. We would round to 4000 since 3870 is closer to 4000 than 3000.

29.
   a. $6 \cdot 7200 + 13 \cdot 360 + 4 \cdot 20 + 2 \cdot 1 = 47{,}962$

30.
   a.

32.
   a.

34.
   a.

36.
   a. $3 \times 5^3 + 0 \times 5^2 + 1 \times 5^1 + 2 \times 5^0 = 382$

39. The student could make ten:
$8 + 7 = 8 + (2 + 5) = (8 + 2) + 5 = 10 + 5 = 15$
or use doubles
$8 + 7 = (1 + 7) + 7 = 1 + (7 + 7) = 1 + 14 = 15$

43.
   a. $y + 3 = 22$
   $y = 22 - 3$
   $y = 19$

44.
   b. $n - 15 = 18$
   $n - 15 + 15 = 18 + 15$
   $n = 18 + 15$
   $n = 33$

46. Answers vary.
   b. Manual put 4 apples in a bucket. Then there were 9 apples in the bucket. How many apples were in the bucket?

50.
   a. 3 ones + 7 ones = 10 ones = 1 ten, 0 ones. The one above the 6 represents the 1 ten from regrouping the ones.

55.
   a.
   Step 1. Build the minuend 345.

   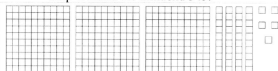

   Step 2. Take away 3 ones, then take away 2 tens.

   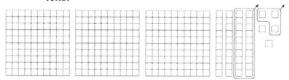

Step 3. That leaves 3 hundreds, 2 tens, and 2 ones.

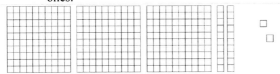

Then 345 − 23 = 322.

56. Answers vary, these are only examples to help show the ideas.
    b. Here it's easier because they are all positive and you could pick a number and add it to 323 to get another term (based on the fact families). 323 + 462 = 785. So it must be true that 785 − 323 = 462.

61. a. $432_{five} \quad \begin{array}{r} \overset{1}{3}04_{five} \\ +123_{five} \\ \hline 432_{five} \end{array}$

62. a. $131_{five} \quad \begin{array}{r} \overset{2\ 10}{\cancel{3}}04_{five} \\ -123_{five} \\ \hline 131_{five} \end{array}$

Chapter 2 Test

1.
   a. True, since 3 is an element of the set.
   b. False, since {3} is an element of the set, but 3 is not an element of the set.

2.
   a. There is a one-to-one correspondence between $\{4, \{0, 1\}\}$ and $\{1, 2\}$: $4 \leftrightarrow 1$, $\{0, 1\} \leftrightarrow 2$. So the cardinality of $\{4, \{0, 1\}\}$ is 2.
   b. There is a one-to-one correspondence between $\{\{5, 2\}, \{1\}, \varnothing\}$ and $\{1, 2, 3\}$: $\{5, 2\} \leftrightarrow 1$, $\{1\} \leftrightarrow 2$, $\varnothing \leftrightarrow 3$. So the cardinality of $\{\{5, 2\}, \{1\}, \varnothing\}$ is 3.
   c. The set is finite. $\{4, 2\} \leftrightarrow 1$, $\{2, 1\} \leftrightarrow 2$, $\{2, 4, 6, 8, ...\} \leftrightarrow 3$. So the cardinality of $\{\{4, 2\}, \{2, 1\}, \{2, 4, 6, 8, ...\}\}$ is 3.

4.
   b. $A \cap B = \{2, 3, 4\}$
   c. $A - B = \{5\}$

5.
   b. 4 students liked all three subjects.
   d. 3 students liked only geometry.

   f. 2 students liked only probability and measurement.

7. The terms in order are: digits, numeral, number.

10.
    a. $300{,}000 + 40{,}000 + 5{,}000 + 600 + 70 + 8$ is the expanded form.
    b. $3 \cdot 100{,}000 + 4 \cdot 10{,}000 + 5 \cdot 1000 + 6 \cdot 100 + 7 \cdot 10 + 8 \cdot 1$ is the expanded form with multiplication.
    c. $3 \cdot 10^5 + 4 \cdot 10^4 + 5 \cdot 10^3 + 6 \cdot 10^2 + 7 \cdot 10^1 + 8 \cdot 10^0$ is the expanded form with exponents.

11. Using the number line model:
    a. 57 to the nearest ten is 60 because it is closer to 60 than to 50 as seen on the number line.

14.
    a. In base ten it is 3124 while the Mayan is:

15.
    a. $2345 \div 360 = 6 \text{ R} 185$; $185 \div 20 = 9 \text{ R} 5$. There are 6 groups of 360, 9 groups of 20, and 5 ones. So 2345 in Mayan is:

    b. $754{,}321 \div 60^3 = 3 \text{ R} 106{,}321$; $106{,}321 \div 60^2 = 29 \text{ R} 1921$; $1921 \div 60^1 = 32 \text{ R} 1$. This is 3 groups of 60^3, 29 groups of 60^2, 32 groups of 60, and 1 group of 1. So 754,321 in Babylonian is:

18.
    a. subtraction property of equality.
    b. addition and subtraction are inverse operations.
    c. addition property of equality.
    d. addition and subtraction are inverse operations.

19.
    b. The definition of subtraction:
       $k - 26 = 84$
       $k = 84 + 26$
       $k = 110$

21. The student would check his work by addition to see if the sum is 65: 38 + 33 = 71. So his answer of 33 is incorrect.

23.
   a. There is a 1 over the 5 because we had 6 ones and 8 ones which is 14 ones. This can be regrouped as 1 ten and 4 ones, and the 1 group of ten is written above the 5 in the tens place to be added to the other groups of tens later.

24.
   a. We have 6 ones, and we need to take away 7 ones. There are not enough ones, so we regrouped 1 ten as 10 ones. This leaves 7 tens instead of 8 tens (since 8 tens, 6 ones = 7 tens, 16 ones).

25.
   a. $567 + 215 \approx 570 + 220 = 790$ (rounded to the nearest ten). Or, $567 + 215 \approx 600 + 200 = 800$ using front-end rounding.
   c. $73 - 37 \approx 70 - 40 = 30$ (rounded to the nearest ten)

Chapter 3 - Section 3.1

2.
  a. The set model of 5×3 is

  • • •   • • •   • • •   • • •   • • •

3.
  a. The number line model of 5×3 is:

  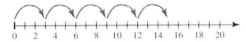

6.
  b. The table shows a 3-by-4 array of possibilities. Because $3 \times 4 = 12$, the student can order 12 possible lunches.

	S	L	O	G
T	TS	TL	TO	TG
P	PS	PL	PO	PG
H	HS	HL	HO	HG

7.

The diagram suggests Kyle has six times as many coins as Luis.

9.
  Use a pattern to solve this problem:
  $456 \times 10^0 = 456$ 3 digits $(0 + 3 = 3)$
  $456 \times 10^1 = 4560$ 4 digits $(1 + 3 = 4)$
  $456 \times 10^2 = 45{,}600$ 5 digits $(2 + 3 = 5)$
  456×10^{78} has $78 + 3 = 81$ digits.

11. If each digit is odd, then there are 5 different values a digit could be. There would be 5 choices for the first digit, 5 for the second, and 5 for the third. So there would be $5 \times 5 \times 5 = 125$ different 3 digit numbers consisting entirely of odd digits.

13.
  a. measurement model $4 \times 12{,}000 = 48{,}000$
  b. set model $4 \times 8 = 32$
  c. Cartesian product model $7 \times 4 = 28$
  d. comparison model $5 \times 7 = 35$
  e. array model $2 \times 200 = 400$

14.
  b. We will use the closure property to view $a + b$ as a whole number, view $c + d$ as the sum of two numbers c and d, then use the distributive property of multiplication over addition (on the right side):
  $(a + b)(c + d) = (a + b)c + (a + b)d$
  $= ac + bc + ad + bd$
  $= ac + ad + bc + bd$

  The formula $(a+b)(c + d) = ac + ad + bc + bd$ is also called the FOIL method:
  First = ac, Outer = ad, Inner = bc, Last = bd

17.
  b. If we know that $a \times b = 0$, then at least one of a or b must be 0.

18.
  b. To prove that $a = b$, we can use the distributive property of multiplication over subtraction. If we know that $a \geq b$ and $5 \times a = 5 \times b$, then we can use the subtraction property of equality to get $5 \times a - 5 \times b = 0$, which can be rewritten using the distributive property of multiplication over subtraction: $5(a - b) = 0$. Since 5 is not 0 and the product is 0, then $a - b = 0$, which means $a = b$.

21.
  a. Count by 3: 3, 6, 9, 12, 15, so $5 \times 3 = 15$

23.
  a. $700 = 7 \times 100$

25. If the letters can be repeated, then there are $26^5 = 11{,}881{,}376$ possible different codes by the fundamental principle of counting.

27. There are 5 boxes, and each box contains 8 oranges. Write an expression involving multiplication that tells how many oranges there are altogether.

32. Answer vary, but the solutions should look like word problems in Problem 31. Here are some possibilities:
  a. An example of the set model: "Lacie has 7 teacups and will put 9 jellybeans in each one. How many jellybeans does she need?"
  b. An example of the measurement model: "Sean jogs 7 miles per day. How many miles does he jog in 9 days?"
  c. An example of the array model: "A classroom is set up in 7 rows with 9 seats in each row. How many seats are in the classroom?"
  d. An example of the Cartesian product model: "A café has 7 different flavors of coffee and 9 different colored cups. How many ways can a customer select a cup of coffee?"

e. An example of the comparison model: "John has 9 coins, and Dewey has 7 times that many. How many coins does Dewey have?"

34.
   a. $7(4 + 8a) = 7 \cdot 4 + 7 \cdot (8a) = 28 + (7 \cdot 8)a$
                                          $= 28 + 56a$

36.
$10 \times (11 \times 9) = (11 \times 9) \times 10$ commutative prop.
                    $= 11 \times (9 \times 10)$ associative prop.

39.
   a. $3000 \times \$35 = \$105{,}000$
   b. $3000 \times 35 = (1000 \times 3) \times 35$
                    $= 1000 \times (3 \times 35)$
                    $= 1000 \times 105$
                    $= 105{,}000$

40.
   a. $4 \times 3 + 5 \times 3 = (4 + 5) \times 3$.

41.
   a. $5 \times 13 = 5 \times (10 + 3) = 5 \times 10 + 5 \times 3 = 50 + 15 = 65$

42.
   a.
$4 \times 125 = 4 \times (25 \times 5) = (4 \times 25) \times 5 = 100 \times 5 = 500$

45. Answers vary. For example:
   a. Mary saved her weekly allowance to buy a sewing kit for $32. She saved the same amount for 4 weeks. What is her weekly allowance? Write an equation that represents the situation.

49. $(3^2)^3 = 9^3 = 729$ and $3^{2 \times 3} = 3^6 = 729$
$(5^4)^2 = 625^2 = 390{,}625$ and $5^{4 \times 2} = 5^8 = 390{,}625$
$(2^3)^4 = 8^4 = 4096$ and $2^{3 \times 4} = 2^{12} = 4096$

These examples support the conjecture $(a^m)^n = a^{m \times n}$.

52.
   a. $8^{12} = (2^3)^{12} = 2^{3 \times 12} = 2^{36}$

53. Answers vary.
   a. Associative property of multiplication:
$15 \times 20 = 15 \times (2 \times 10) = (15 \times 2) \times 10$
                                $= 30 \times 10 = 300$

56. Multiplying powers of ten along with the commutative properties lets us write:
$600 \times 40 = (6 \times 4) \times (100 \times 10)$
                $= 24 \times 1000 = 24{,}000$.

59.
   a. $5a + 3a = 8a$, and $3a$ is a counting number. By the definition of *less than*, $5a < 8a$.

61.
   a. 5 and 6 belong to {5, 6, 7} but 5×6 does not belong to {5, 6, 7}.

62.
   a. $2 \times 2 = 4$ which is in the set. $6 \times 4 = 24$ which is in the set. $6 \times 8 = 48$ which is in the set. Therefore, by inductive reasoning, the product of two even numbers is an even number. The set of even numbers is closed under multiplication.

Section 3.2

1.
   a. A division sentence answering the question "how many groups" would be $18 \div 6 = 3$ since there are 18 objects in 3 groups of 6.

2.
   a. A division sentence answering the question "how many objects in each group" would be $18 \div 3 = 6$ since there are 18 objects in 3 groups of 6.

5.

$77 - 5 = 72$
$72 \div 4 = 18$
$3 \times 18 + 5 = 59$
Matthew has 18 coins.
Courtney has 59 coins.

7.
   a.

$20 \div 4 = 5$
Marco has 5 coins.

   b. Let M represent the number of coins Marco has. Then Tyson has 4M coins.
$4M = 20$
$M = 5$
Marco has 5 coins.

9.
   a. 122 − 8 = 114. The average blue whale weighs 114 more tons than a Hummer.
   b. 122 ÷ 8 = 15 R2. The average blue whale weighs 2 more tons than 15 times as much as a Hummer.

12. We know that $y = 3q + 2$.
   a.
   $(y+1) \div 3 = (3q + 2 + 1) \div 3 = (3q + 3) \div 3$
   $= (q + 1) \times 3 \div 3 = q + 1$

13.
   a. There are 5 groups of 3 and 15 objects altogether, so $5 \times 3 = 15$.
   b. There are 15 objects and 3 objects in each of the 5 groups, so $15 \div 3 = 5$
   c. There are 15 objects and 5 groups of 3, so $15 \div 5 = 3$.

17. Answers vary.
   a. A teacher has 16 pieces of candy. He wants to split it evenly among 3 students. How many pieces of candy would be left over?
   b. A teacher has 16 pieces of candy. He wants split it evenly among 3 students. How many more pieces of candy would he need to give all of it away?
   c. A teacher has 16 pieces of chocolate. He wants to put 3 in each bag to give to his colleagues. How many bags of chocolate can he make?

20.
   a. $3 \times n = 12$ because the student would "know" n equals 4.
   b. Division can be used to solve for d in $d \times 543 = 225,345$ (by dividing: $d = 225,345 \div 543$).

23. $100,000 \div 15 = 6666$ R10 and $1000 \div 45 = 2222$ R10. The candidate had about 2222 to 6666 times the normal level of dioxin, which suggests the candidate was poisoned. Answers vary, but a sensational headline could read "Presidential candidate poisoned! Dioxin levels over 6000 times the normal amount."

26. fair share model
   a. There are 4 groups of 2 dots each.

   $9 \div 4 = 2$ R1

27. repeated subtraction model
   a. Two groups of 4 dots will be removed.

   $9 \div 4 = 2$ R1

28. Using repeated subtraction:
   a. 24 − 8 = 16
      16 − 8 = 8
      8 − 8 = 0
      So $24 \div 8 = 3$.

29.
   a. $n \div 3 = 5$
      $n = 3 \times 5$
      $n = 15$

31.
   a. Because $0 \le r < d$ for remainder r and divisor d, then $0 \le r < 7$, meaning there are 7 possible values (0, 1, 2, 3, 4, 5, and 6).

33. Solve using algebra:
   a. Let t equal the number of marbles Tony has. Then we know that Marco has $3t + 2$ but also has 41 marbles, so we can solve for t with
      $3t + 2 = 41$
      $3t = 39$
      $t = 13$
      So Tony has 13 marbles.

34.
   a.

   453 − 6 = 447, 447 ÷ 3 = 149
   Peter has 149 trading cards.

38. Verify the division:
   a. $2 \times 8 + 1 = 16 + 1 = 17$, so $17 \div 2 = 8$ R1
   b. $1 \times 7 + 5 = 7 + 5 = 12$, so $12 \div 7 = 1$ R5
   c. $3 \times 5 = 15$, so $15 \div 3 = 5$

39. Fact families:
   a. $28 \div 7 = 4$; $28 \div 4 = 7$; $4 \times 7 = 28$; $7 \times 4 = 28$.

40.
   a. $23^8 \div 23^3 = a$
      $23^{8-3} = a$
      $23^5 = a$

43. No; $1010101 \div 10101 = 100$ R1, and the nonzero remainder means $10101 \times a = 1010101$ does not have a whole number solution.

48. Let *n* represent the number of pieces of candy. Then we know that $n \div 5 = q$ R3 and $n \div 7 = p$ R4. We can write these as $n = q \times 5 + 3$ and $n = p \times 7 + 4$. Then $7p + 4 = 5q + 3$. The $7p + 1 = 5q$. So this means that $7p + 1$ is a multiple of 5. Choosing values of *p*, we can find the smallest one possible: $7 \times 1 + 1 = 8$; $7 \times 2 + 1 = 15$. So the smallest value is when $p = 2$, meaning $n = 2 \times 7 + 4$, $n = 14 + 4 = 18$. So the minimum number of pieces of candy is 18.

49.
  a. Subtract: Saturn is 1,277,127,510 km farther from the Sun than Earth.
  b. Divide: The distance is 80,344,390 km more than 9 times the distance of the Earth to the Sun, or Saturn is about 10 times as far from the Sun as Earth.
  c. Multiplicative reasoning is more appropriate because it is easier to comprehend "about 10 times" as far than comprehend "1,277,127,510 km farther."

50.
  a. $288 - 48 = 240$. An 85-pound student burns 240 more calories per hour moving boxes than watching TV.
  b. $288 \div 48 = 6$. An 85-pound student burns 6 times as many calories per hour moving boxes as watching TV.

55.
  a. There is only one whole number where $a \div 14 = 1$ and that is $a = 14 \times 1 = 14$.
  b. There is only one whole number where $35 \div b = 1$ and that is $b = 35 \div 1 = 35$.
  c. If we know that *a* and *b* are whole numbers and $a \div b = 1$, we can conclude that $a = 1 \times b$, meaning $a = b$ and $b \neq 0$.

56.
  a. $7,838,400 \div 28,400 = 276$. There were 276 audience members. The repeated subtraction model is more appropriate because the number of groups is unknown.
  b. $550,000 \div 20 = 27,500$. Each TV station was fined $27,500. The fair-share model of division is more appropriate, because the number of objects per group is unknown.

60. We know that $y = 5q + 3$.
  a. $(y + 4) \div 5 = (5q + 3 + 4) \div 5 =$
    $(5q + 7) \div 5 = (5q + 5 + 2) \div 5 =$
    $((q + 1) \times 5 + 2) \div 5 = (q + 1)$ R 2.
    So $(y + 4) \div 5 = (q + 1)$ R2

63.
  a.
  $18 \times n = 90$
  $18 \times n \div 18 = 90 \div 18$
  $n = 90 \div 18$
  $n = 5$

  b.
  $18 \times n = 90$
  $n = 90 \div 18$
  $n = 5$

Section 3.3

1.
Step 1. Build four copies of 37.

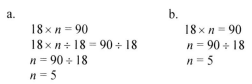

Step 2. Regroup the 28 ones as 2 tens and 8 ones.

Step 3. Regroup the 14 tens as 1 hundred and 4 tens. No more regrouping is needed.

There are 1 hundred, 4 tens, and 8 ones. So $4 \times 37 = 148$.

2.
  a. $863 \div 4 \approx 800 \div 4 = 200$
  b. **Step 1.** Build 863 using base ten blocks.

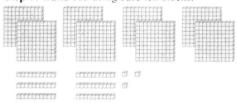

Step 2. Separate the hundreds and tens into four equal-sized groups.

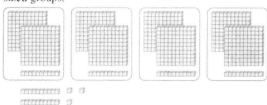

Step 3. Regroup the remaining 2 tens as 20 ones. Then 2 tens plus 3 ones equals 23 ones.

Step 4. Divide the 23 ones equally among the four groups. There are 3 ones left over.

Each group gets 2 hundreds, 1 ten, and 5 ones, with 3 remaining. Then $863 \div 4 = 215$ R3.

5. Use the Guess and Check strategy:
   a. 36 and 7 ($6 \cdot 7 = 42$ and $30 \cdot 7 = 210$)

7. Applying the standard multiplication algorithm to 234×56 requires the simpler problems 234×50 and 234×6.

9.
   a. The sum of the partial quotients is $400 + 30 + 60 + 3 = 493$. Then
   $3456 \div b = 493$ R5
   $3456 = b \times 493 + 5$
   $3456 - 5 = b \times 493$
   $3451 = b \times 493$
   $b = 7$
   The divisor is 7.
   b. An equation involving division is:
   $3456 \div 7 = 493$ R5.
   c. Checking our work,
   $7 \times 493 + 5 = 3451 + 5 = 3456$.

13.
   a. We are regrouping 2 hundreds, 3 tens as 23 tens.
   b. We are regrouping 1 ten, 7 ones as 17 ones.

15.
   a. $456 \times 82 = 37{,}392$
   b. $49 \times 36 = 1764$

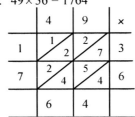

16.
   a. $(4x^2 + 8x + 3)(6x + 5) = 24x^3 + 68x^2 + 58x + 15$

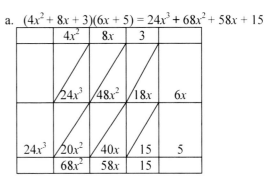

17. The product equals the sum of the partial products.

18. The quotient equals the sum of the partial quotients.

20. Monique is calculating the simpler product 30×7, which equals $(3 \times 7) \times 10$. The zero anticipates multiplying the product 3×7 by 10.

24. Using partial products:
   a.
   ```
      127
    ×  45
       35
      100
      500
      280
      800
     4000
     5715
   ```

25. Answers vary. For example, 654 and 23.

29.
   a. The quotient is the sum of the partial quotients: $15 + 10 + 3 + 1 = 29$. Knowing the dividend, quotient, and remainder, we can find the divisor:
   $a = b \times q + r$
   $2329 = b \times 29 + 67$
   $2262 = b \times 29$
   $b = 78$
   The divisor is 78.
   b. An equation is $2329 \div 78 = 29$ R67.
   c. $29 \times 78 + 67 = 2262 + 67 = 2329$.

31.
   a. The quotient is the sum of the partial quotients: $21 + 5 + 120 + 32 = 178$. Knowing the divisor, quotient, and remainder, we can find the dividend:

$a = b \times q + r$
$a = 43 \times 178 + 35$
$a = 7689$
The dividend is 7689.
   b. An equation is $7689 \div 43 = 178$ R35.
   c. $43 \times 178 + 35 = 7654 + 35 = 7689$.

32. Estimates and partial quotients vary.
   a. Estimate: $1841 \div 23 \approx 1800 \div 20 = 90$.

   ```
   23 )1 8 4 1
        -2 3 0      10
         1 6 1 1
         -4 6 0     20
          1 1 5 1
          -9 2 0    40
            2 3 1
           -2 3 0   10
               1
   ```
   $1841 \div 23 = 80$ R1

34.
   a. $4304 \div 30 \approx 4200 \div 30 = 140$

   ```
        143
   30)4304
       -30
       130
      -120
       104
        -90
         14
   ```

   $4304 \div 30 = 143$ R14

40. Answers vary.
   a. $40 \times 50 = 2000$

41. The front end estimate of 34 is 30, and $30 \times y = 2400$ leads to $y = 80$. Any whole number that rounds down to 80 will work: 80, 81, ..., 89.

43. Find a range based estimate:
   a. Low: $24 \times 365 \Rightarrow 20 \times 300 = 6000$;
      High: $24 \times 365 \Rightarrow 30 \times 400 = 12,000$.

44. a. 98; $15 \times 100 = 1500$ and $15 \times 2 = 30$, so $1500 - 30 = 1470$.

47. Answers vary.
   a. $2 \times 4 \times 6 \times 5 = 2 \times 5 \times 6 \times 4 = 10 \times 24 = 240$

48. Answers vary.
   a. $8 \times 12 = 8 \times (10 + 2) = 80 + 16 = 96$

53. One block has a surface area of 128 square inches. The wall is 6 feet (72 inches) high and 120 feet (1440 inches) long, for a surface area of $72 \times 1440 = 103,680$ square inches. Then an approximation of the number of blocks needed would be $103,680 \div 128 = 810$. So at a minimum, Gary will need 810 blocks.

56. In the end, there were 42 baskets which had an average of 20 items each, which is $42 \times 20 = 840$ items. The manager added 32 items to get this total, so the original amount was $840 - 32 = 808$ items.

59. All numbers could be a partial quotient or partial product, so the table will be:

Number	300	320	11	40	43
/, \, or X	X	X	/	X	/

60. The two simpler problems are $234_{five} \times 40_{five}$ and $234_{five} \times 2_{five}$ because
$234_{five} \times 42_{five} = 234_{five} \times (40_{five} + 2_{five})$
$= 234_{five} \times 40_{five} + 234_{five} \times 2_{five}$

61. The two numbers being multiplied are 241_{six} and 24_{six} because
$241_{six} \times 24_{six} = 241_{six} \times (4_{six} + 20_{six})$
$= 241_{six} \times 4_{six} + 241_{six} \times 20_{six}$

62.
   a. $3_{five} \times 4_{five} = 22_{five} = 2_{five}$ longs, 2_{five} units. The 2 above the 4 represents the 2_{five} longs when 22_{five} is regrouped as 2_{five} longs, 2_{five} units.
   b. $43_{five} \times 4_{five} = 332_{five}$, as follows:

   ```
     3 2
     43_{five}
   ×  4_{five}
   ───────
    332_{five}
   ```

66.
   a. The student regrouped 11_{five} flats and 2_{five} longs as 112_{five} longs.
   b. The 112 represents 112_{five} longs.

69. Given that $3324_{five} \div 12_{five} = 231_{five}$ R2_{five}, find:
   a. $3320_{five} \div 12_{five}$. The dividend here is 4 less than the original, but there are only 2 left over. We'll need to reduce the quotient to get more in the remainder. But reducing the quotient from 231_{five} to 230_{five} increases the remainder by one group of 12_{five}, so the new (improper) equation would be:

$3324_{five} \div 12_{five} = 230_{five} \text{ R}14_{five}$. We can remove 4 from the remainder and dividend to get the result we need:
$3320_{five} \div 12_{five} = 230_{five} \text{ R}10_{five}$.

b. $3323_{five} \div 12_{five}$. The dividend is exactly one less than the given equation, so the result just involves removing one from the remainder to balance it out. $3323_{five} \div 12_{five} = 231_{five} \text{ R}1_{five}$.

c. $3330_{five} \div 12_{five}$. The dividend here is 1 more than the original, so this 1 would fit in the remainder giving us the equation $3330_{five} \div 12_{five} = 231_{five} \text{ R}3_{five}$.

d. $3332_{five} \div 12_{five}$. The dividend here is 3 more than the original, so these 3 would fit (barely) in the remainder giving us the equation $3332_{five} \div 12_{five} = 231_{five} \text{ R}10_{five}$.

72.
  a. The quotient is the sum of the partial quotients: $100_{five} + 120_{five} + 10_{five} + 2_{five} + 1_{five} = 233_{five}$.
  b. Knowing the dividend, quotient, and remainder, we can find the divisor using $a = b \times q + r$. Just subtract the remainder from the dividend, then divide the result by the quotient and you'll be left with the divisor:
  $a = b \times q + r$
  $23020_{five} = b \times 233_{five} + 3_{five}$
  $23012_{five} = b \times 233_{five}$
  $b = 44_{five}$
  The divisor is 44_{five}.

Chapter 3 Review

2.
  c. measurement model

3. Answers vary.
  b. An invalid IP address would be 1.3.5.2.5 (it has 5 blocks instead of 4) or 555.555.555.555 (can only be between 0 and 255).

5. Answers vary.
  b. Joey had 3 boxes of books. Each box had the same number of books. There were 24 books in all. Write a number sentence for the problem.

9. Use skip counting.
  a. Count out multiples of 5 until you've listed 4 of them: 5, 10, 15, 20. So $4 \times 5 = 20$.

10. Use the distributive property.
  a. $15 \times 11 = 15 \times (10 + 1) = 15 \times 10 + 15 \times 1 = 150 + 15 = 165$

13.
  a. $68^2 = 4624$

14.
  a. If you know $4 \times 6 = 24$, then you can find that $40 \times 6 = (10 \times 4) \times 6 = 10 \times (4 \times 6) = 10 \times 24 = 240$

19.
  a. $8^4 = (2^3)^4 = 2^{3 \times 4} = 2^{12}$

22.
  a. The repeated subtraction model since he knows how many pennies are in each stack.
  b. $47 \div 8 = 5 \text{ R}7$, so there will be 7 pennies left over.

26. Remind the student that multiplication and division are inverse operations.
$$\begin{aligned} n \div 7 &= 84 \\ n \div 7 \times 7 &= 84 \times 7 \\ n &= 84 \times 7 \\ n &= 588 \end{aligned}$$

32.
  c. $77 = 4q + 1$, then $76 = 4q$, then $q = 19$.

36. Answers vary.
  b. Mrs. Dugger had 14 raffle tickets. She wanted to split them equally among 3 students who won them. How many raffle tickets does each student get?

40. The two simpler problems are 456×30 and 456×2, because $456 \times 30 + 456 \times 2 = 456 \times (30 + 2) = 456 \times 32$.

44.
  a.
  $$\begin{aligned} abc \times ef &= abc \times (10e + f) \\ &= abc \times 10e + abc \times f \\ &= 35{,}220 + 2348 = 37{,}568 \end{aligned}$$

48.
a.
$$\overset{1}{2}13_{\text{five}}$$
$$\times\ 23_{\text{five}}$$
$$\overline{1144}$$
$$\underline{4310}$$
$$11004_{\text{five}}$$

52.
a. $36 \times 198 \approx 36 \times 200 = 7200$

57. We are regrouping 1 hundred, 3 tens as 13 tens.

Chapter 3 Test

2.
a. $38 \times 400 = 38 \times (4 \times 100) = (38 \times 4) \times 100 = 152 \times 100 = 15{,}200$
b. $42 \times 103 = 42 \times (100 + 3) = 42 \times 100 + 42 \times 3 = 4200 + 126 = 4326$

4. a.

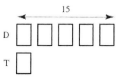

$15 \div 5 = 3$. Tony has 3 coins.
b. Let T represent the number of coins that Tony has, so Daniel has 5T coins.
$5T = 15$
$T = 15 \div 5$
$T = 3$
Tony has 3 coins.

7.
a. $3 \cdot 3 \cdot 3 \cdot 3 = 3^4$
b. $7^{43} \cdot 7^{82} = 7^{43+82} = 7^{125}$

8.
a. $10^{12} = (2 \cdot 5)^{12} = 2^{12} \cdot 5^{12}$
b. $20^{25} \cdot 10^{31} = (2^2 \cdot 5)^{25} \cdot (2 \cdot 5)^{31} = (2^2)^{25} \cdot 5^{25} \cdot 2^{31} \cdot 5^{31} = 2^{2 \times 25 + 31} \cdot 5^{25+31} = 2^{81} \cdot 5^{56}$

10. a. repeated subtraction: $14 \div 4 = 3\ \text{R}\ 2$
b. fair share: $14 \div 3 = 4\ \text{R}\ 2$

11. a. division property of equality
b. multiplication and division are inverse operations
c. multiplication property of equality
d. multiplication and division are inverse operations

13.
a.
$n \times 24 = 120$
$n = 120 \div 24$
$n = 5$

b.
$n \times 24 = 120$
$n \times 24 \div 24 = 120 \div 24$
$n = 120 \div 24$
$n = 5$

14.
a. 8, 16, 24. Then $3 \times 8 = 24$.
b. $12 \times 8 = (10 + 2) \times 8 = 10 \times 8 + 2 \times 8 = 80 + 16 = 96$

16. The student can check the work for $44 \div 6 = 7\ \text{R}\ 3$ by checking if $6 \times 7 + 3$ equals 44. $6 \times 7 + 3 = 42 + 3 = 45$. This does not equal the dividend 44, so the students made an error.

20. Answers vary.
a. Lisa had 32 beads. She divided them into equal-sized piles. There were 8 beads in each pile. How many piles were there?
b. The gardener has 22 flowers. He wanted to plant 4 flowers in each row. How many rows of flowers did he plant?

22.
a. The distributive property of multiplication over addition allows this:
$234 \times 56 = 234 \times (50 + 6) = 234 \times 50 + 234 \times 6$
b. The two numbers are 482 and 57 because $482 \times 57 = 482 \times (50 + 7) = 482 \times 50 + 482 \times 7$.
c. The two simpler problems required by the standard algorithm are 942×30 and 942×8 because $942 \times 38 = 942 \times (30 + 8) = 942 \times 30 + 942 \times 8$.

23.
c. Lisa is calculating the simpler product 234×500, which equals $(234 \times 5) \times 100$. The two zeros anticipate multiplying 234×5 by 100.

25.
c. The quotient is the sum of the partial quotients: $1000 + 100 + 10 + 10 + 5 + 3 = 1128$. Knowing the dividend, quotient, and remainder, we can find the divisor:
$a = b \times q + r$
$46263 = b \times 1128 + 15$
$46248 = b \times 1128$
$b = 41$
The divisor is 41.

Chapter 4 – Section 4.1

1.
   a.

a	4	7	10	13	16	19
$a \div 3$	1 R 1	2 R 1	3 R 1	4 R 1	5 R 1	6 R 1

4.
   a. The diagram representation of $26 \div 3 = 8$ R 2 would be:

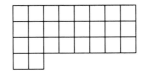

   b. The shaded blocks represent the remainder:

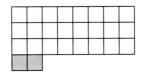

   c. 3 is not a divisor of 26 because each row (or column) does not have the same number of cells or boxes.

5. Each box can include only one color marble, so each box must have a number of marbles that divides both 35 and 20. 5 is a divisor of both, so there would be 5 marbles in each box, making 7 boxes of yellow marbles and 4 boxes of green marbles, for a total of 11 boxes of marbles.

7. Since it must be less than 40, we could just list the possibilities. He had a number that was a multiple of 5 with 2 left over (2, 7, 12, 17, 22, 27, 32, 37) and also was a multiple of 7 with 4 left over (4, 11, 18, 25, 32, 39). 32 is the only number in both lists, so Mr. Smith had 32 students in his class.

8.
   a. You can tell in this case that 6 does not divide the number because 6 divides 12 but does not divide 32.
   b. You can tell in this case that 6 does not divide the number because 2 does not divide the number.

11.
   a. No, 3 does not divide 1000.
   b. For any values of a and b, 3 divides the left-hand side of the equation $9a + 36b$, since $9a + 36b = 3(3a + 12b)$, but 3 does not divide the right-hand side of the equation 1000. So there are no values of a and b that solve the equation.

13. 30 divides n, so $n = 30k$ for some whole number k. Then $n = 15(2k)$. We know $2k$ is a whole number, so 15 divides n.

14.
   a. The remainder is 1 since 2 divides $2 \cdot 3 \cdot 5$.

15.
   b. $678 \to 6 \cdot 7 \cdot 8 = 336 \to 3 \cdot 3 \cdot 6 = 54 \to 5 \cdot 4 = 20 \to 2 \cdot 0 = 0$. There were 4 steps, so the step number is 4.

17.
   d. Once you know the first 12 digits, you can find the weighted sum of the first 12 digits. Then the check digit is chose to make a valid weighted sum as shown above. If it was chosen anything but last, there would be the chance that it would not make a valid ISBN.

22.

ISBN	9	7	8	1	4	0	2	0	1	n	5	8	2
Weight	1	3	1	3	1	3	1	3	1	3	1	3	1
Product	9	21	8	3	4	0	2	0	1	$3n$	5	24	2

Weighted sum = $9 + 21 + 8 + 3 + 4 + 0 + 2 + 0 + 1 + 3n + 5 + 24 + 2 = 3n + 79$. Solve to find n so that the weighted sum is divisible by 10; so $n = 7$, since 21 is the only multiple of 3 that has ones digit of 1.

25. 4 divides 32 and 4 does not divide 7.
By Theorem 3(a), 4 does not divide $32 + 7$.

26. 3 divides 42 and 3 divides 12.
By Theorem 2(b), 3 is a divisor of $42 - 12$.

28. Answers vary.
   c. Neither 6 nor 7 are factors of 45.

30. c. divisibility rules

33. Answers vary
   a. If $a = 6$ and $b = 4$, then $a + b = 10$, and 5 divides $a + b$.
   b. If $a = 6$ and $b = 7$, then $a + b = 13$, and 5 does not divide $a + b$.

35. Suppose a is a divisor of b and b is a divisor of c. Then $b = ap$ and $c = bq$ for some whole numbers p and q. Then $c = bq = (ap)q = a(pq)$. We know pq is a whole number by the closure property of multiplication of whole numbers, meaning a is a divisor of c.

38.
  a. a_n is divisible by 3 whenever the digit sum is divisible by 3. Since the sequence is blocks of 1's, if there are groups of 3 ones, then it will be divisible by 3; this happens when n is a multiple of 3.
  b. a_n will never be divisible by 5 because the ones digit will always be 1.
  c. a_n will never be divisible by 6 because the ones digit will always be 1, meaning that a_n is not divisible by 2. Since a_n is not divisible by 2, it cannot be divisible by 6.

39.
  c. $a^6 = a^{3 \times 2} = (a^3)^2$, so a^6 is a perfect square.

41.
  a. The divisibility test for 4 relies on the 2 rightmost digits of the number. $92 \div 4 = 23$, so 4 divides 92, thus 4 divides 2,296,492.
  b. The divisibility test for 4 relies on the 2 rightmost digits of the number. $58 \div 4 = 14 \text{ R} 2$, so 4 does not divide 58, therefore 4 does not divide 20,858.

45.
  a. $3 \times 8 = 24$ and the only common factor of 3 and 8 is 1. Then 24 divides a number if and only if 3 and 8 divide the number.
  b. The last 3 digits (016) are divisible by 8, and $1 + 2 + 5 + 0 + 1 + 6 = 15$, which is divisible by 3, so 125,016 is divisible by 24.
  c. The last 3 digits (608) are divisible by 8, and $7 + 4 + 6 + 0 + 8 = 25$, which is not divisible by 3, so 74,608 is not a multiple of 24.

47.
  a. $abcd = ab \cdot 100 + cd$. Any number n can be written in the form $n = k \cdot 100 + xy$, where k is a whole number and x and y are digits. We know 25 divides 100. This suggests the following divisibility rule: 25 divides the whole number n, if and only if, 25 divides the number formed by two rightmost digits of n.
  b. No, since 89 is not divisible by 25.
  c. Yes, since 75 is divisible by 25.

49.
  a. There are possible three values for b: 4, 2 and 1. The sequence is arithmetic with common difference of 4, so dividing by 4 or any factor of 4 would produce the same remainder each time.

52.
  a. The only common factor of 4 and 5 is 1. A rule is 20 divides a number n if and only if 4 divides n and 5 divides n.
  b. Example: Does 20 divide 7180? 4 divides 7180 and 5 divides 7180. Then 20 divides 7180. Check: $7180 \div 20 = 359$.

54.
  a. Dividing by 2:

a	2	6	10	14	18	22
$a \div 2$	1 R0	3 R0	5 R0	7 R0	9 R0	11 R0

  The remainders form the repeating sequence 0, 0, 0, 0, ... or all the remainders are 0.

56.
  b. False. 14 is not a divisor of 21 but 7 is a divisor of 21.

57.
  a. No. $10^4 \cdot 3^5 \cdot 17^8$ cannot be multiplied by a whole number to get to to $10^2 \cdot 3^4 \cdot 17^6$, as there are too many 10's, 3's and 17's in $10^4 \cdot 3^5 \cdot 17^8$.
  b. Yes. $U = 5^{12} \cdot 2^4 \cdot 11^{600} = (5^1 \cdot 2^2)(5^{11} \cdot 2^2 \cdot 11^{600}) = 20(5^{11} \cdot 2^2 \cdot 11^{600})$.

Section 4.2

3. This is not a prime factorization because 8 is not a prime number.

4.
  a. $\text{GCF}(90, 144) = 2 \cdot 3^2 = 18$
  b. $\text{LCM}(90, 144) = 2 \cdot 3^2 \cdot 2^3 \cdot 5 = 2^4 \cdot 3^2 \cdot 5 = 720$

5.
  a. The Venn representation of $72 = 2^3 \cdot 3^2$ and $132 = 2^2 \cdot 3 \cdot 11$ is:

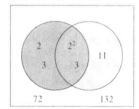

10. 16 flowers. We need a number that will divide both 48 and 32. Obviously 1 works, but let's find GCF(32,48) to possibly obtain a larger number. $32 = 2^5$ and $48 = 2^4 \cdot 3$, so the GCF of 32 and 48 is $2^4 = 16$. So put 16 flowers of the same type in each container. This would correspond to 2 containers of daffodils and 3 containers of tulips.

11. 60 pieces. Since he could have 1, 2, 3, 4, or 5 friends, then we need to find a number N that is divisible by each. The prime factorization helps make this easier. Since our number N is divisible by the primes 2, 3, and 5, and $4 = 2^2$, we could just write the largest exponent and find the solution quickly: $N = 2^2 \cdot 3 \cdot 5 = 60$. This is the LCM of 1, 2, 3, 4, and 5.

14.
   a. 60 years. The LCM of 12 and 30 is 60 using lists: 12, 24, 36, 48, 60, 72 ... and 30, 60, 90, ...
   b. 420 years. The LCM of 12, 30 and 84 can be found quickly using $84 = 12 \cdot 7$. Since 7 does not divide 12 or 30, then we just need to multiply $\text{LCM}(12,30) = 60$ by 7 to find $\text{LCM}(12,30,84) = 60 \cdot 7 = 420$. The list or prime factorization method would show the same result.
   c. 420 years. The Earth has a revolving time of 1 year, and since it has no new factors, then LCM(1, 12, 30, 84) = $\text{LCM}(12,30,84) = 60 \cdot 7 = 420$. The list or prime factorization method would show the same result.

16. The LCM of two numbers cannot exceed the product of the two numbers. 102 is larger than 3×17.

20. The left-hand side of the equation has 5 as one of the prime factors. The right-hand side of the equation does not have 5 as one of the prime factors. The prime factors of $5n$ do not match the prime factors of 24. The prime factorization of a number is unique, so $5n \neq 24$ for any whole number n.

21. No. The number n would have to have the form $n = p^a \cdot q^b$ for some prime numbers p and q and counting numbers a and b. Then a and b would have to satisfy the equation $(a+1) \cdot (b+1) = 5$. $(a+1) \cdot (b+1)$ is a product of two counting numbers greater than 1, and this would imply 5 is a composite number. However, we know 5 is a prime number.

23.
   a. The Mersenne Prime that filled 12 newspaper pages had 378,632 digits.
   $378,632 \div 12 = 31,552$ R8 tells us that there are about 31,552 digits per page.

   $7,235,733 \div 31,522 = 229$ R17,195. Then 230 pages would be required.

24.
   c. Multiples of 42: 42, 84, 126, $\boxed{168}$, 210, ...
   Multiples of 24:
   24, 48, 72, 96, 120, 144, $\boxed{168}$, ... These two lists agree on 168, which is the LCM.

32. Since p is greater than 2 and is also prime, then we know p must be odd (2 is the only even prime). Adding 1 to an odd number makes it even, and divisible by 2. So $p + 1$ must be composite.

34. $\text{GCF}(a,b) \cdot \text{LCM}(a,b) = a \cdot b$, so the only time the $\text{LCM}(a,b) < a \cdot b$ is when the GCF is greater than 1. If a and b have at least one common prime factor, then $\text{GCF}(a,b) > 1$ and $\text{LCM}(a,b) < a \cdot b$.

36. g. composite

37. We know 1 and 42 are factors, and since 42 is even, we can find a few more. $42 \div 2 = 21$, so 2 and 21 are also factors. $42 \div 3 = 14$, so 3 and 14 are factors. 4 and 5 do not divide 42, but since 2 and 3 divide 42, we know 6 will too. $42 \div 6 = 7$, so 6 and 7 are factors. 7 is already in the list, so we have found all the factors of 42. The factors of 42 in the order of discovery are: 1, 42, 2, 21, 3, 14, 6, 7.

39.
   a. Because the prime factorization is $3^5 \cdot 2^7$, there are $(5+1)(7+1) = 6 \cdot 8 = 48$ factors.

40. Because the prime factorization is $3^5 \cdot 7^4 \cdot 11^2$, there are $(5+1)(4+1)(2+1) = 6 \cdot 5 \cdot 3 = 90$ factors.
   a. Of the 90 factors, there are only 3 prime factors which are the base values in the prime factorization.

41. Because this is a factorization and not a prime factorization, we must first find the prime factorization. $20^5 \cdot 3^4 \cdot 13^2 = (2^2 \cdot 5)^5 \cdot 3^4 \cdot 13^2 = 2^{10} \cdot 5^5 \cdot 3^4 \cdot 13^2$. Since the prime factorization is $2^{10} \cdot 5^5 \cdot 3^4 \cdot 13^2$, there are $(10+1)(5+1)(4+1)(2+1) = 11 \cdot 6 \cdot 5 \cdot 3$ = 990 factors.
   a. Of the 990 factors, there are only 4 prime factors which are the base values in the prime factorization.

45. The GCF is 24, which is the greatest common factor. However, every number that divides 24 will also divide the two numbers *a* and *b*. Since there are 8 divisors of 24, then there are 8 total common factors of *a* and *b*.

47. Since the only factorization of a prime number is $p \cdot 1$, a prime number must a counting number that has exactly one proper divisor.

48.
   b. We can conjecture that the number of factors in a perfect square is odd.

53. Answers may vary, but the concept is the same for all.
   a. If we want a number with 12 divisors and 2 are prime, then we'll need to write the number in the form $a = p^n \cdot q^b$, with primes of *p* and *q*, and counting numbers *n* and *b*. Then using the idea developed in this section, the number of factors will be $(n+1) \cdot (b+1)$. For this problem, any primes will work, and as long as the exponents are chosen to satisfy $(n+1) \cdot (b+1) = 12$, the number *a* will be acceptable. Factorizations of twelve include $2 \cdot 6$ and $3 \cdot 4$. Notice 1 and 12 are also factors, but having 1 in the factorization means $n = 0$ or $b = 0$, which would make only one prime factor. So the exponents are 1 and 5 or 2 and 3. Some values for *a* are $a = 7^1 \cdot 13^5$ or $a = 5^2 \cdot 23^3$ or $a = 2^3 \times 5^2$.

54. Start with the number 217. Write it as 217,217. Now divide by 7 to get $217,217 \div 7 = 31,031$. Divide next by 11: $31,031 \div 11 = 2821$. Finally, divide by 13 to get $2821 \div 13 = 217$. We seem to get back to the original number we started with. But why did this happen? Start with a representation of any three digit number: "*abc*". When we write the number twice in a row, it now looks like "*abcabc*". Then we divide by 7, 11, and 13. But $7 \cdot 11 \cdot 13 = 1001 = 1000 + 1$. $abc \cdot 1001 = abc \cdot (1000 + 1) =$ $abc \cdot 1000 + abc = abcabc$. This shows that any 3 digit number repeated to form a 6 digit number is divisible by 1001. Dividing *abcabc* by 7, 11, and 13 is like dividing $abc \cdot 7 \cdot 11 \cdot 13$ by 7, 11, and 13. The result is *abc*.

56. There are no numbers that make this a prime number because $12b + 4 = 4(b+3)$, making *n* divisible by 4, which means that *n* cannot be a prime number (it must have factors of at least 1, 2, 4, and *n*).

57. Using the largest exponents on each prime factor:
   a. $\text{LCM}(a,b) = 2^4 \cdot 3^4 \cdot 5^4 \cdot 7^2 \cdot 11$.

60. Answers vary.
   a. Since $\text{GCF}(a,b) = 2^3 \cdot 5^2 \cdot 7^4 \cdot 11$, any two numbers with the GCF above will work. For example, $a = 2^5 \cdot 5^2 \cdot 7^8 \cdot 11$ and $b = 2^3 \cdot 5^4 \cdot 7^4 \cdot 11$.

61.
   a. $\text{LCM}(a,b) = 1$ and $\text{LCM}(a,b) \geq \text{GCF}(a,b)$, then $1 \geq \text{GCF}(a,b)$. LCM and GCF are whole numbers, so $1 = \text{GCF}(a,b)$.
   b. $\text{GCF}(a,b) \cdot \text{LCM}(a,b) = a \cdot b$ and $\text{GCF}(a,b) = 1$, so $1 \cdot \text{LCM}(a,b) = a \cdot b$. Then $\text{LCM}(a,b) = a \cdot b$.

63. Using repeated division, choose a prime number and divide. Continue choosing prime numbers to divide; the divisors and remainder that you obtain are the factors.
   a.

   $2\overline{)6}$ 3

   $2\overline{)12}$

   $2\overline{)24}$

   $2\overline{)48}$

   Then $48 = 2^4 \times 3$

Section 4.3

1.
   d. 3; three white chips.

2.
   a. Place 3 black chips and then put 2 more black chips next to them (adding the 2 black to the 3 black) resulting in 5 black chips. This is why $-3 + -2 = -5$ as seen here:

3. Place 3 black chips which represents the minuend of -3.

   We need to subtract 4 (or 4 white chips). To do this put in zero pairs in order to have 4 white and 4 black chips.

This still represents –3, and now we can take away the subtrahend, leaving behind 7 black chips illustrating why –3 – 4 = –7.

7. Start each problem at 0 facing the positive direction. Walk forward for positive and backwards for negative, but if you subtract, then turn around and face the opposite direction.
   a. Start by walking backwards 3 units, then (because it is subtraction) turn around to face the negative direction, then walk forward 4 more units to end at –7: –3 – 4 = –7.

9.
   d. With 4 negatives and 2 positives, there are 2 zero pairs. This leaves 2 negatives behind, or the integer –2.

10.
   d. Start with 4 negative charges. Since we are subtracting 3 negative charges, we don't need to include any zero pairs. Just subtract the 3 negatives and it leaves one negative charge: – 4 – –3 = –1

14. Since the busboy works 7 hours and makes $6 per hour, then he would typically make $7 \cdot \$6 = \$42$ on a shift if he didn't break any dishes. Let d represent the number of broken dishes. Then $42 - 2d = 34$. Solving for d, we get $d = 4$. The busboy broke 4 dishes during his shift.

15.
   a. All of these integers can be written without a negative sign, making them all non-negative integers.
   b. All of these numbers are less than or equal to zero, so they are all non-positive integers.

16. The reasons are:
   a. addition property of equality
   b. associative property of addition

17. The reasons are:
   a. definition of Subtraction
   b. definition of Subtraction

18.
   b. Similar to part (a), if the sign of $n \cdot n$ is positive, then multiplying by n one more time will create $n \cdot n \cdot n$. However, if n is negative, then this will be negative and if n is positive, then this is positive.

21. The reasons are:
   a. addition property of equality
   b. additive identity property
   c. associative property of addition
   d. additive inverse property
   e. additive identity property

22.
   a. No. For example, 5 – 3 is a whole number, but 3 – 5 (which equals –2) is not a whole number. Since it is not true for all whole numbers, the set of whole numbers is not closed under subtraction.
   b. Yes. For any two integers a and b, we can subtract by adding the opposite: $a - b = a + - b$. Because the opposite of an integer is an integer, then we have turned subtraction of two integers into addition of two integers, and the set of integers is closed under addition. The set of integers is closed under subtraction because every difference of integers is also an integer.

26. The next equation would be:
   a. (the first factor is decreasing by 1 each time, and the product is decreasing by 4 each time) $-1 \cdot 4 = -4$

27.
   b. This set would have all positive integers and all negative integers, but would not have 0. $I^- \cup I^+ = \{..., -3, -2, -1, 1, 2, 3, ...\}$. This could also be written as all integers removing just 0: $I^- \cup I^+ = I - \{0\}$.

29. The temperature was –3°F this afternoon, but once it got dark it dropped 7° F more degrees. What was the temperature after dark?

31. First, evaluate $32 \div 8$ to get 4. Note that 32 and –8 have different signs, so the quotient is negative. Then $32 \div -8 = -4$.

32. k. sign rules   l. debts and assets

33. d. –8   e. 0

35. If you plot both numbers on the number line, then the number on the left is less than the number on the right. Negative integers are always to the left of zero and positive integers are always to the right of zero. So any negative integer is always to the left of any positive integer. So any negative integer is less than any positive integer.

36

38.
a.

$n - {-3}$	=	8	Given
n	=	$8 + {-3}$	Definition of Subtraction
n	=	5	Simplification

39.
b. $-2 - 8 = -2 + {-8} = -10$

42. To use patterns, start with some facts the student will know and work to the result desired.
a.
$2 + 4 = 6$
$1 + 4 = 5$
$0 + 4 = 4$
$-1 + 4 = 3$
$-2 + 4 = 2$
$-3 + 4 = 1$

45. Using the sign rules:
a. 3 and –9 have different signs, so $3 \times {-9}$ is negative. $3 \times 9 = 27$, so $3 \times {-9} = -27$.

46. Using the sign rules:
a. 36 and –9 have different signs, so $36 \div {-9}$ is negative. $36 \div 9 = 4$, so $36 \div {-9} = -4$.

47. Using properties of multiplication:
a.

$-2n - 6n$	=	$-2n + {-6n}$	Subtraction by adding the opposite.
	=	$(-2 + {-6})n$	Distributive Property of Multiplication over Addition
	=	$-8n$	Simplification

50.
b. $-|{-7} - {-2}| = -|{-5}| = -5$

55. The distance between integers a and b is $|a - b|$.
a. $|{-3} - 5| = |{-8}| = 8$

59. The number of terms in an arithmetic sequence with first term a, last term b (with $a < b$), and common difference $d = 1$ is given by the formula $b - a + 1$:
a. $-2 - {-9} + 1 = -2 + 9 + 1 = 7 + 1 = 8$
b. $31 - {-12} + 1 = 31 + 12 + 1 = 43 + 1 = 44$
c. $30 - a + 1 = 31 - a$

61.
a. The integer n must satisfy the following equations: $n - 2 = 8$ or $n - 2 = -8$. The solutions to these are $n = 10$ or $n = -6$.

Chapter 4 Review

3.
a. To represent $23 \div 4 = 5$ R3, we can build a rectangle with 5 columns of 4 each with 3 left over.

4.
c. No, 5 is not a multiple of 15 (15 is a multiple of 5).

6.
a. Yes; 2 divides 712 and 2 divides 930, so we know that 2 will divide 712 + 930.

8.
a. No; 5 does not divide 1532 but 5 does divide 120, so we know that 5 will not divide 1532 – 120.

9.
b. From this equation, we can tell that both 36 and 285 do not divide 10,300, and that attempting the division with either number will leave a remainder of 4.

12. Suppose 4 is a factor of n. Then $n = 4k$ for some counting number k. Then $n + 3 = 4k + 3$. We know 4 divides $4k$ and 4 does not divide 3, so 4 does not divide $4k + 3$. But $n + 3 = 4k + 3$, so 4 does not divide $n + 3$.

15. The divisibility test for 4 is that the last 2 digits must be divisible by 4.
a. Since the last 2 digits are divisible by 4, then the largest digit must be 9 making the number 43,908.

17. In order for the three digit number *abc* to be divisible by 3, $a + b + c$ must be divisible by 3. Since addition is commutative, then other numbers divisible by 3 would be *acb*, *bca*, or *bac*. Other possibilities are *aabbcc*, *abcabc*, and *bacbacbac*.

19.
   a. Since $22 = 2 \cdot 11$ and the greatest common factor of 2 and 11 is 1, then a divisibility test for 22 is to check both 2 and 11: 22 divides n if and only if 2 divides n and 11 divides n.
   b. 2 and 11 both divide 2706, so 22 will divide 2706.
   c. 2 divides 928, but 11 does not divide 928. So 22 does not divide 928.

22.
   b. The number has 2 factors, so we know the number is a prime number. In this case, the prime number is 6829.

24. 24 and 8 have a common factor of 2, so we can write $B = 24 \cdot A + 8 = 2(12 \cdot A + 4)$. Then B has at least three factors (1, 2, and B). Then B is a composite number for all values of A.

29. In order to find all factors with repeated division, we must increase the divisor each time.
   a. For the number 60, we know 1 and 60 are factors. Moving to 2, we see that 2 and 30 are factors. Continuing to 3 we get 3 and 20 as factors. Testing 4 we see that 4 and 15 are factors. Checking 5 we see that 5 and 12 are factors. With 6, it is clear that 6 and 10 are factors. 7 is not a factor, nor are 8 and 9. So all the factors of 60 in the order of discovery are: 1, 60, 2, 30, 3, 20, 4, 15, 5, 12, 6, 10.

31.
   a. $2^3 \cdot 5^6 \cdot 11^2$ is a prime factorization, so the number of factors is $(3+1) \cdot (6+1) \cdot (2+1) = 4 \cdot 7 \cdot 3 = 84$. From the factorization, we see 3 prime factors. Of the 84 factors, there are $84 - 3 - 1 = 80$ composite factors (since 1 is neither prime nor composite).

35.
   a. $21^2 \leq 451$, so we need to check the primes from 2 to 21. 451 is not divisible by 2, 3, or 5 by the divisibility tests, but 451 is divisible by 11. So 451 is composite.

37. Using the list method.
   a. Factors of 12 are 1, 2, ③, 4, 6, and 12; factors of 15 are 1, ③, 5, and 15. So $GCF(12, 15) = 3$. Multiples of 12 are 12, 24, 36, 48, ⑥⓪, 72, 84, 96, ... while multiples of 15 are 15, 30, 45, ⑥⓪, 75, ... So $LCM(12, 15) = 60$.

38. Using the prime factorization method, we choose the largest exponent on each prime to find the LCM.
   a. $LCM(a, b) = 2^3 \cdot 3^4 \cdot 5^2 \cdot 7 \cdot 13$.

39. Using the prime factorization method, we choose the largest exponent on each prime to find the GCF.
   a. $GCF(a, b) = 2^3 \cdot 5^2$.

40. Using the Euclidean Algorithm to find the GCF, we divide and find remainders, and then repeat.
   a. $GCF(346, 68) = 2$ because:
   $346 = 68 \cdot 5 + 6$
   $68 = 6 \cdot 11 + 2$
   $6 = 2 \cdot 3 + 0$

45. Model $-5 + 3$ using:
   a. The chip model. Start by putting in 5 black chips and then add 3 white chips to this. Ignore zero pairs and you will be left with 2 black chips showing $-5 + 3 = -2$.

49.
   a. $4 - -7 = 4 + 7 = 11$.

52.
   a. The factors have different signs so the product will be negative. $2 \cdot 8 = 16$, so $2 \cdot (-8) = -16$.

53.
   a. The dividend and divisor have different signs so the quotient will be negative. $18 \div 6 = 3$, so $18 \div -6 = -3$.

Chapter 4 Test

2.
   b. This shows that 57 is not divisible by 17, and that dividing 57 by 17 gives a remainder of 6.

5.
   a. Yes. The prime factorization of 175 is $175 = 5^2 \cdot 7$. We can express B in the form $B = (5^2 \cdot 7)(5^{212} \cdot 7^{42} \cdot 11^{10})$, or $B = 175 \cdot (5^{212} \cdot 7^{42} \cdot 11^{10})$

7.
  a. Because $18 = 2 \cdot 9$ and $\text{GCF}(2,9) = 1$, we know that "a number is divisible by 18 if and only if that number is divisible by 2 and 9."

13. $12^3 \cdot 5^4 \cdot 11^{15}$ is not a prime factorization, so we must first put it in prime factorization.
$12^3 \cdot 5^4 \cdot 11^{15} = (3 \cdot 2^2)^3 \cdot 5^4 \cdot 11^{15} = 3^3 \cdot 2^6 \cdot 5^4 \cdot 11^{15}$.
Then $12^3 \cdot 5^4 \cdot 11^{15}$ has $(3+1)(6+1)(4+1)(15+1) = (4)(7)(5)(16) = 2240$ factors.

22.
  a. $-5 - 3 = -5 + -3 = -8$.

24. Sign rules for multiplication and division are if the signs are the same, the result is positive and it is negative if the signs are different.
  a. 2 and –8 have different signs, so the product is negative. $2 \times 8 = 16$, so $2 \times (-8) = -16$.

26. Answers vary.
  a. Marco borrowed $7 from his parents to buy lunch each week for 5 weeks. How much money did he owe them altogether?

Chapter 5 – Section 5.1

1.
  a. The 3 tells you that the collection consists of three equal-sized parts.
  b. The 5 tells you that each equal-sized part in the collection is called one-fifth.
  c. Answers vary, but here is one example.
  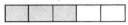

5.
  b. Set model representation:
  $$\frac{3}{5} \qquad \frac{6}{10}$$
  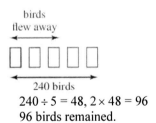

  In this model, the whole is represented by

6.
  b. As a mixed number, this value is $1\frac{2}{8}$.

7. One circle represents a whole unit, and each circle is partitioned into 7 parts.
  a. There are 4 whole circles for 28 parts with 3 more parts for a total fraction of $\frac{31}{7}$.

9. Sketch a unit representing the 240 birds and let three-fifths of it represent the birds that flew away.

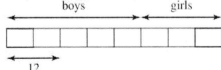

  $240 \div 5 = 48, 2 \times 48 = 96$
  96 birds remained.

10. We know that five-eighths of the students in class are boys, meaning three-eighths are girls.

  Since three of the blocks represent girls, the boys have two more blocks (5 total). Because there are 12 more boys than girls, these two blocks represent the 12 more boys. That means there are $12 \div 2 = 6$ students in each block. Eight total blocks gives us $8 \cdot 6 = 48$. There are 48 total students (30 boys and 18 girls).

14. For Molly, the 'whole' is the entire 3 by 4 unit array:

We shade 1 of the 4 columns to make 1/4.

For Ben, since he sees it as $\frac{3}{4}$, and 3 pieces are shaded, then each piece would represent $\frac{1}{4}$. His model for the whole would be:

That way we can shade 1 of the 4 pieces to make $\frac{1}{4}$

and the 3/4 would be modeled by

17.
  a. Six circles represent three-fifths. We split the circles into three groups.
  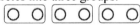
  Each of these groupings is one-fifth of the whole. So one possible diagram of one-fifth is:

  Another possible diagram is obtained by shading two of six circles:
  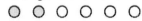

  b. Since this rectangle represents three-fifths, we can split it into 3 equal-sized parts and each one will represent one-fifth.

  c. Parition the number line from 0 to $\frac{3}{5}$ into three equal-sized pieces. The number $\frac{1}{5}$ is shown on the number line.

19. $\dfrac{440}{760} = \dfrac{44}{76} \approx \dfrac{45}{75} = \dfrac{3}{5}$

21.
a. Since the weight is 290 grams, we can set up a table to find the appropriate fraction of 290:

Fraction of 290	$\frac{1}{4}$	$\frac{1}{5}$	$\frac{1}{6}$
Weight of cake	72.5 g	58 g	48.33 g

58 grams is closest to the reference amount 55 grams, so the serving size is 58 g ($\frac{1}{5}$).

23. We could compare each to another number and perhaps that would help. In this situation, we know that $\frac{3}{7} < \frac{1}{2}$ and $\frac{1}{2} = \frac{4}{8} < \frac{5}{8}$ which means $\frac{3}{7} < \frac{1}{2} < \frac{5}{8}$, so $\frac{3}{7} < \frac{5}{8}$.

25.
a. $\frac{8}{5}$, since a pizza can be divided into equal-sized pieces.
b. $8 \div 5 = 1$ R3, since pencils cannot be divided into equal-sized pieces.

28. Without any calculations: $\frac{2}{7} < \frac{1}{2}$ and $\frac{1}{2} < \frac{4}{5}$, so $\frac{2}{7} < \frac{4}{5}$.

31.
a. He had $35 and spent two-fifths on a gift card. We will draw a picture of the $35 as a rectangle and let two-fifths represent the money spent on a gift card.

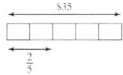

Then $35 \div 5 = 7$. Then $7 \cdot 2 = 14$. He spent $14 on a gift card.

b. Start by representing the 72 grams of sugar with a rectangle partitioned into 8 equal-sized pieces. Then 5 of the pieces represent the amount spilled (five-eighths).

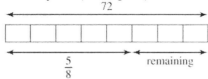

$72 \div 8 = 9$, $3 \times 9 = 27$.
He has 27 grams of sugar left.

33. Using the Fundamental Law of Fractions, we can multiply the numerator and denominator by the same number to obtain an equivalent fraction. Answers will vary as there are infinitely many equivalent fractions.
a. $\frac{3}{4} = \frac{3 \cdot 2}{4 \cdot 2} = \frac{6}{8}$, $\frac{3}{4} = \frac{3 \cdot 3}{4 \cdot 3} = \frac{9}{12}$

34. We can divide the numerator and denominator by the same number to obtain an equivalent fraction. Answer will vary if a different divisor is chosen.
a. $\frac{24}{32} = \frac{24 \div 2}{32 \div 2} = \frac{12}{16}$, $\frac{24}{32} = \frac{24 \div 4}{32 \div 4} = \frac{6}{8}$

39.
a. $1543 \div 35 = 44\frac{3}{35}$, $1543 \div 35 = 44$ R3.

40.
a. $41 \div 12 = 3$ R5, $41 \div 12 = 3\frac{5}{12}$.

46.
a. 3 boxes represent one batch and we have 4 boxes, so we have $\frac{4}{3}$ batches or $1\frac{1}{3}$ batches.

47.
b. $\frac{435}{821} \approx \frac{400}{800} = \frac{1}{2}$

49.
c. Unit fractions can be ordered based on their denominators. If $0 < m < n$, then $\frac{1}{n} < \frac{1}{m}$.

50.
c. Fractions with common denominators can be ordered based on their numerators. If $0 < a < b$, then $\frac{a}{c} < \frac{b}{c}$.

53.
b. Since $\frac{1}{8} < \frac{1}{7}$, we know that $\frac{12}{8} < \frac{12}{7}$.

54. Answers vary greatly.
b. $\frac{7}{11} = \frac{21}{33}$ and $\frac{8}{11} = \frac{24}{33}$, so two fractions between $\frac{7}{11}$ and $\frac{8}{11}$ are $\frac{22}{33}$ and $\frac{23}{33}$.

57. We can use Theorem 5.1 and the concept of cross products there to solve.
a. $\frac{634}{787} < \frac{360}{n}$
$634n < 787 \cdot 360$
$n < \frac{787 \cdot 360}{634} = \frac{283,320}{634} = 446\frac{556}{634}$
So the largest value of n would be $n = 446$.

61.
c. Since he has read two-fifths of the book and has to read 30 more pages to finish the book, then we could represent the entire book as 5 equal-sized parts. The 30 pages would represent how much more he has to read.

$30 \div 3 = 10$, $5 \times 10 = 20$.
There are 50 pages in the book.

62.
a. 540 mph. Splitting the indicated airspeed 450 into five equal pieces would make each piece $450 \div 5 = 90$, so the actual speed would be 90 miles per hour faster than 450 mph.

64.
a. Since the sequence is arithmetic, there must be a common difference. $1\frac{2}{5} = \frac{7}{5}$, which is $\frac{4}{5}$ larger than $\frac{3}{5}$. The next three terms must be: $3\frac{4}{5}$, $4\frac{3}{5}$, and $5\frac{2}{5}$.

Section 5.2

2. Draw a large rectangle to represent one whole, and partition it into eighths since $\frac{2}{8} = \frac{1}{4}$. This diagram shows $\frac{1}{4} + \frac{3}{8} = \frac{5}{8}$.

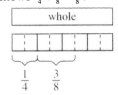

4. To subtract with different denominators, create a whole unit partitioning it into equal-sized parts corresponding to the LCM of the two denominators. In this case, create a whole unit that is partitioned into eighths. $\frac{6}{4} = \frac{12}{8} = 1\frac{4}{8}$, and the 12 shaded squares represent $\frac{6}{4}$. We then remove $\frac{5}{8}$ by putting an "x" in 5 boxes. The diagram shows that $\frac{6}{4} - \frac{5}{8} = \frac{7}{8}$.

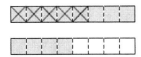

5.
a. Estimate the sum: $6 + 4 = 10$. The actual sum is $5\frac{3}{4} + 3\frac{2}{3} = 5\frac{9}{12} + 3\frac{8}{12} = 8\frac{17}{12} = 9\frac{5}{12}$.

6.
a. Estimate the difference: $9 - 4 = 5$. The actual difference is $8\frac{5}{6} - 3\frac{3}{4} = 8\frac{10}{12} - 3\frac{9}{12} = 5\frac{1}{12}$.

10.
a. $\frac{1}{2} + \frac{1}{4} = \frac{2}{4} + \frac{1}{4} = \frac{3}{4}$.
b. $\frac{1}{2} + \frac{1}{4} + \frac{1}{8} = \frac{4}{8} + \frac{2}{8} + \frac{1}{8} = \frac{7}{8}$.
c. If we try one more to confirm the pattern, we notice that $\frac{1}{2} + \frac{1}{4} + \frac{1}{8} + \frac{1}{16} = \frac{8}{16} + \frac{4}{16} + \frac{2}{16} + \frac{1}{16} = \frac{15}{16}$. The pattern seems to be adding unit fractions with denominators that double each time, and the sum is a fraction with the largest denominator as the denominator and one less than that as the numerator. Using this pattern we can conjecture that $\frac{1}{2} + \frac{1}{4} + \frac{1}{8} + \ldots + \frac{1}{1024} = \frac{1023}{1024}$.

11. If $\frac{x}{y}$ and $\frac{u}{v}$ are additive inverses of $\frac{a}{b}$, then it must be true that $\frac{x}{y} + \frac{a}{b} = 0$ and $\frac{u}{v} + \frac{a}{b} = 0$ from the definition of additive inverses. Since $\frac{x}{y} + \frac{a}{b} = 0$, we can subtract $\frac{a}{b}$ from both sides to get $\frac{x}{y} = -\frac{a}{b}$. Similarly for $\frac{u}{v} + \frac{a}{b} = 0$ we get $\frac{u}{v} = -\frac{a}{b}$. Because $\frac{x}{y}$ and $\frac{u}{v}$ are both equal to the same expression, then $\frac{x}{y} = \frac{u}{v}$.

15.
a. Because there is only one pizza, we can see if this is possible by adding the fractions: $\frac{3}{5} + \frac{4}{7} = \frac{21}{35} + \frac{20}{35} = \frac{41}{35}$. This is larger than 1 pizza, so this situation is not possible. We could also (in this case) notice that $\frac{3}{5} > \frac{1}{2}$ and $\frac{4}{7} > \frac{1}{2}$, so adding these would be greater than 1.
b. In this situation, $\frac{5}{8} + \frac{2}{7} = \frac{35}{56} + \frac{16}{56} = \frac{51}{56}$. Since $\frac{51}{56} < 1$ this situation is possible.

16.
a. Using the definition of subtraction:

Equation	Justification
$n - \frac{3}{8} = \frac{5}{6}$	Original equation
$n = \frac{5}{6} + \frac{3}{8}$	Definition of Subtraction
$n = \frac{20}{24} + \frac{9}{24}$	Fundamental Law of Fractions
$n = \frac{29}{24}$	Definition of Addition of Rational numbers
$n = 1\frac{5}{24}$	Simplification

b. Using the fact that addition and subtraction are inverses:

Equation	Justification
$n - \frac{3}{8} = \frac{5}{6}$	Original equation
$n - \frac{3}{8} + \frac{3}{8} = \frac{5}{6} + \frac{3}{8}$	Addition Property of Equality
$n = \frac{5}{6} + \frac{3}{8}$	Addition and Subtraction are inverse operations
$n = \frac{20}{24} + \frac{9}{24}$	Fundamental Law of Fractions
$n = \frac{29}{24}$	Definition of Addition of Rational numbers
$n = 1\frac{5}{24}$	Simplification

c. Using the basic properties only:

Equation	Justification
$n - \frac{3}{8} = \frac{5}{6}$	Original equation
$\left(n - \frac{3}{8}\right) + \frac{3}{8} = \frac{5}{6} + \frac{3}{8}$	Addition Property of Equality
$\left(n + -\frac{3}{8}\right) + \frac{3}{8} = \frac{5}{6} + \frac{3}{8}$	Subtraction by Adding the Opposite Property
$n + \left(-\frac{3}{8} + \frac{3}{8}\right) = \frac{5}{6} + \frac{3}{8}$	Associative Property of Addition
$n + 0 = \frac{5}{6} + \frac{3}{8}$	Additive Inverse Property
$n = \frac{5}{6} + \frac{3}{8}$	Additive Identity Property
$n = \frac{20}{24} + \frac{9}{24}$	Fundamental Law of Fractions
$n = \frac{29}{24}$	Definition of Addition of Rational numbers
$n = 1\frac{5}{24}$	Simplification

20. Writing as a sum of distinct unit fractions:
   a. $3 \div 8 = \frac{3}{8} = \frac{2+1}{8} = \frac{2}{8} + \frac{1}{8} = \frac{1}{4} + \frac{1}{8}$.
   b. $5 \div 8 = \frac{5}{8} = \frac{4+1}{8} = \frac{4}{8} + \frac{1}{8} = \frac{1}{2} + \frac{1}{8}$.
   c. $5 \div 6 = \frac{5}{6} = \frac{3+2}{6} = \frac{3}{6} + \frac{2}{6} = \frac{1}{2} + \frac{1}{3}$.

22. A fraction is a collection of equal sized pieces. The sum of two fractions with common denominators means we are combining two collections of the same sized pieces. The sum then becomes just counting the number of these equal sized pieces.

25. Answers vary.
   a. For example, let $\frac{1}{4}$ be one fraction and $\frac{5}{7} - \frac{1}{4} = \frac{13}{28}$ be the other. Then $\frac{13}{28} + \frac{1}{4} = \frac{5}{7}$.

30.
   a. Using the LCM of the denominators to add:
   $\frac{5}{6} + \frac{7}{8} = \frac{20}{24} + \frac{21}{24} = \frac{41}{24} = 1\frac{17}{24}$.

31.
   a. Using the LCM of the denominators to subtract:
   $\frac{7}{8} - \frac{5}{6} = \frac{21}{24} - \frac{20}{24} = \frac{1}{24}$.

32.
   b. Using addition and subtraction as inverse operations:

Equation	Justification
$\frac{4}{3} + n = \frac{9}{5}$	Original equation
$\frac{4}{3} + n - \frac{4}{3} = \frac{9}{5} - \frac{4}{3}$	Subtraction Property of Equality
$n = \frac{9}{5} - \frac{4}{3}$	Addition and Subtraction are inverse operations
$n = \frac{27}{15} - \frac{20}{15}$	Fund. Law of Fractions
$n = \frac{7}{15}$	Simplification

33.
   a. Using the definition of subtraction:

Equation	Justification
$\frac{3}{4} - y = \frac{5}{6}$	Original equation
$\frac{3}{4} = \frac{5}{6} + y$	Definition of Subtraction
$\frac{3}{4} - \frac{5}{6} = y$	Definition of Subtraction
$\frac{9}{12} - \frac{10}{12} = y$	Fundamental Law of Fractions
$-\frac{1}{12} = y$	Definition of Subtraction of Rational numbers

34. The distance between $\frac{3}{4}$ and $1\frac{11}{20}$ is $1\frac{11}{20} - \frac{3}{4} = \frac{31}{20} - \frac{15}{20} = \frac{16}{20}$. We can split $\frac{16}{20}$ into 4 equal-sized parts: $\frac{16}{20} \div 4 = \frac{16}{20} \times \frac{1}{4} = \frac{4}{20} = \frac{1}{5}$. Then $A = \frac{3}{4} + \frac{1}{5} + \frac{1}{5} = \frac{15}{20} + \frac{4}{20} + \frac{4}{20} = \frac{23}{20} = 1\frac{3}{20}$.

42.
   c. $\frac{-55}{16} = \frac{-48-7}{16} = \frac{-48}{16} - \frac{7}{16} = -3\frac{7}{16}$.

47.
   a. $12\frac{1}{5} + 7\frac{5}{6} \approx 12 + 8 = 20$.

49.
   b. $\frac{3}{n} + \frac{4}{5} = 1\frac{8}{35} \Rightarrow \frac{105}{35n} + \frac{28n}{35n} = \frac{43}{35} \Rightarrow$
   $\frac{105+28n}{35n} = \frac{43n}{35n} \Rightarrow 105 + 28n = 43n \Rightarrow$
   $105 = 15n \Rightarrow 7 = n$.

Section 5.3

1.
   a. 4/6 of the whole is marked with /'s, and then of this region 3/5 are marked with \'s. The cross-hatched region marked by ×'s is $\frac{3}{5} \cdot \frac{4}{6} = \frac{12}{30}$.

3.
   a. $\frac{13}{4} \div \frac{5}{7} = \frac{13}{4} \times \frac{7}{5}$

5. $105 + 3 = 108, 108 \div 9 = 12, 2 \times 12 - 3 = 21$. Neal has 21 coins.

6.
   a. $\frac{2}{5}$. Since Kyle has 3 boxes and Hannah has 5 boxes, Kyle has 2 fewer boxes which is equivalent to $\frac{2}{5}$ of Hannah's amount.
   b. $\frac{2}{3}$. Since Hannah has 2 more boxes than Kyle, she has $\frac{2}{3}$ more because 2 boxes is $\frac{2}{3}$ of Kyle's amount.

9.
   a. David has 6 more than $\frac{2}{5}$ of the number of beads that Erin has. The total is 671, so we can draw the following:

   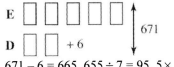

   $671 - 6 = 665$, $655 \div 7 = 95$, $5 \times 95 = 474$, $2 \times 95 + 6 = 196$. David has 196 beads.

   b. Using algebra: Let e represent the number of beads Erin has. David has $\frac{2}{5}e + 6$ beads.
   $\left(\frac{2}{5}e + 6\right) + e = 671$, $\left(\frac{2}{5} + 1\right)e + 6 = 671$,
   $\left(\frac{7}{5}\right)e = 665$, $e = 665 \cdot \left(\frac{5}{7}\right)$, $e = 475$. Erin has 475 beads, and $\frac{2}{5}e + 6 = \frac{2}{5} \cdot 475 + 6 = 196$. David has 196 beads.

11. Suppose $\frac{a}{b}$ and $\frac{c}{d}$ are any two rational numbers. Then $\frac{a}{b} \times \frac{c}{d} = \frac{ac}{bd}$ and $\frac{c}{d} \times \frac{a}{b} = \frac{ca}{db} = \frac{ac}{db} = \frac{ac}{bd} = \frac{a}{b} \times \frac{c}{d}$. Then $\frac{a}{b} \times \frac{c}{d} = \frac{c}{d} \times \frac{a}{b}$.

14. The sunflower grew $5 - 2 = 3$ inches taller, while the tomato plant grew $8\frac{1}{4} - 4\frac{1}{2} = 3\frac{3}{4}$ inches taller. So Linda could be correct. $3 \div 2 = 1\frac{1}{2}$ and $3\frac{3}{4} \div 4\frac{1}{2} = \frac{15}{4} \div \frac{9}{2} = \frac{15}{4} \times \frac{2}{9} = \frac{15}{18}$, which means the sunflower grew $1\frac{1}{2}$ times its original height while the tomato plant grew $\frac{15}{18}$ times its original height. So Barry could be correct.

16.
   b. Using the definition of division:

Equation	Justification
$y \times \frac{5}{6} = 24$	Original expression
$y = 24 \div \frac{5}{6}$	Definition of division
$y = 24 \times \frac{6}{5}$	Using invert and multiply rule
$y = \frac{144}{5}$	Definition of Multiplication of Rational Numbers
$y = 28\frac{4}{5}$	Simplification

17.
   c. Using invert and multiply rule:

Equation	Justification
$y \div \frac{7}{3} = 65$	Original expression
$y \times \frac{3}{7} = 65$	Invert and Multiply Rule
$\left(y \times \frac{3}{7}\right) \times \frac{7}{3} = 65 \times \frac{7}{3}$	Multiplication Property of Equality
$y \times \left(\frac{3}{7} \times \frac{7}{3}\right) = 65 \times \frac{7}{3}$	Associative Property of Multiplication
$y \times 1 = 65 \times \frac{7}{3}$	Multiplication Inverse Property
$y = 65 \times \frac{7}{3}$	Multiplication Identity Property
$y = \frac{455}{3}$	Definition of Multiplication of Rational Numbers
$y = 151\frac{2}{3}$	Simplification

19. Final answers rounded to the nearest hundred.
   a. The number of city miles driven is $\frac{1}{6} \times 12{,}000 = 2000$ miles. The number of gallons of gasoline to drive 2000 miles is $2000 \div 24 \approx 83$ gallons. The number of freeway miles driven is $\frac{5}{6} \times 12{,}000 = 10{,}000$ miles. The number of gallons of gasoline to drive 10,000 miles is $10{,}000 \div 42 \approx 238$ gallons. Then $83 + 238 = 321$ and $321 \times 3.50 = \$1123$. The total cost is approximately $\$1100$.

21.
   a. No. Check $\frac{3}{4} \cdot 105 = \frac{315}{4} = 78\frac{3}{4}$. Since this is less than 85, we cannot say that brand X has fewer calories than brand Y.

22. $\frac{0}{3}$ has no reciprocal because the possible reciprocal, $\frac{3}{0}$, is equivalent to $3 \div 0$ which is undefined. So $\frac{0}{3}$ has no reciprocal.

24. Answers vary. Mrs. Tan's class had $4\frac{1}{3}$ pounds of cookie dough. Suppose one batch requires $\frac{3}{4}$ pound of dough. How many batches of cookie could be made?

29.
   c. $8\frac{3}{4} \div 3\frac{2}{5} = \frac{35}{4} \div \frac{17}{5} = \frac{35}{4} \cdot \frac{5}{17} = \frac{175}{68} = 2\frac{39}{68}$.

30.
   b. Using the definition of division:

Equation	Justification
$\frac{4}{7} \times y = 12$	Original expression
$y = 12 \div \frac{4}{7}$	Definition of division
$y = 12 \cdot \frac{7}{4}$	Using invert and multiply rule
$y = \frac{84}{4}$	Definition of Multiplication of Rational Numbers
$y = 21$	Simplification

34. It helps to draw a picture for some of these problems.
   a. Since there are 3 times as many girls as boys, then we could draw one box to represent the boys and 3 to represent the girls. 3 of the 4

boxes represent the girls, so $\frac{3}{4}$ of the class are girls.

b. Since there are two-thirds as many girls as boys, then make 3 boxes for the boys and 2 for the girls. 2 of the 5 boxes represent the girls, so $\frac{2}{5}$ of the class are girls.

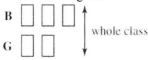

37. Using a diagram, draw a rectangle to represent the milk in the container. Kendra spilled 3/5 of the container, so that means there 2/5 of the milk remained for Oliver.

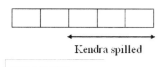

Take the 2 remaining pieces as one whole object and re-draw it as 3 equal sized pieces. This represents the portion Oliver spilled (2 of the blocks) and the 6 ounces originally.

Oliver spilled 2/3 of the remaining milk, and there were 6 ounces remaining. That means Oliver spilled 12 ounces (2 boxes worth), and there were 18 ounces remaining after Kendra spilled the milk. $18 \div 2 = 9$, so each fifth represents 9 ounces of milk. Then $5 \times 9 = 45$. Originally, there were 45 ounces of milk in the container.

39.
a. Using a diagram: George planted 5 more than $1\frac{1}{3}$ times as many trees as Martha.

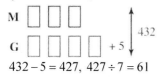

$432 - 5 = 427$, $427 \div 7 = 61$

$4 \times 61 + 5 = 249$

George planted 249 apple trees.

b. Using algebra: Let m equal number of apple trees Martha planted. Then George planted $1\frac{1}{3}m + 5$ apple trees. Then $m + \frac{4}{3}m + 5 = 432$, $\frac{3}{3}m + \frac{4}{3}m = 432$, $\frac{7}{3}m = 427$,

$m = 427 \cdot \frac{3}{7}$, $m = 183$. Martha planted 183 trees, so George planted the rest: $432 - 183 = 249$. We could also check:
$1\frac{1}{3}m + 5 = 1\frac{1}{3} \cdot 183 + 5 = 244 + 5 = 249$.

45. The reasons (in order) are:
a. rule for subtracting fractions
b. rule for multiplying fractions
c. distributive property of multiplication over subtraction.
d. rule for subtracting fractions
e. rule for multiplying fractions

48.
a. The whole is a rectangle partitioned into four equal-sized parts. Then we will draw $\frac{9}{4} = 2\frac{1}{4}$ and create groups of $\frac{5}{4}$:

Each group of $\frac{5}{4}$ requires 5 pieces that are $\frac{1}{4}$ in size. We see that there is one whole group of $\frac{5}{4}$, and 4 pieces left. There is not enough left to make another group, since there are 4 of the 5 required pieces. This shows $\frac{9}{4} \div \frac{5}{4} = 1\frac{4}{5}$.

50.
b. $\frac{2}{3} \cdot 2\frac{3}{5} = \frac{2}{3} \cdot \frac{13}{5} = \frac{26}{15} = 1\frac{11}{15}$. This is not the same as $\frac{8}{5}$, so the student's work is incorrect.

53.
a. $\frac{8}{36} = \frac{2}{9}$. Amelia has 8 fewer books. Compared to Nancy's amount, these 8 would be $\frac{8}{36}$ of Nancy's amount. So Amelia has $\frac{8}{36}$ fewer books than Nancy.

b. $\frac{28}{36} = \frac{7}{9}$. Amelia has 28 books compared to Nancy's 36. The 28 would represent $\frac{28}{36}$ of Nancy's amount. So Amelia has $\frac{28}{36}$ times as many books as Nancy.

c. $\frac{8}{28} = \frac{2}{7}$. Nancy has 8 more books, which would represent $\frac{8}{28}$ of Amelia's total. So Nancy has $\frac{8}{28}$ more books than Amelia.

d. $\frac{36}{28} = 1\frac{2}{7}$. Nancy has the same number of books Amelia does plus 8 more books, which would represent $\frac{36}{28}$ of Amelia's total. So Nancy has $\frac{36}{28}$ times as many books as Amelia.

56.
   a. multiplication property of equality

57.
   a. $4\frac{1}{2} \div \frac{2}{3} = \frac{9}{2} \cdot \frac{3}{2} = \frac{27}{4} = 6\frac{3}{4}$. This means Jenny walked $6\frac{3}{4}$ times as far as Hillary.

60.
   b. The base is –3 and the exponent is 4.

61.
   a. Product rule since there are 2 factors with a base of x.

Section 5.4

1. The ratio would be 4 to 5, 4:5, or $\frac{4}{5}$.

5. Since the ratio is 3 to 7, we can create a table with the number of pens and pencils in rows.

number of pens	3	6	9	12
number of pencils	7	14	21	28

7.
   a. The boxes represent the ratio that Julie has 3 beads for every 5 beads Ann has.

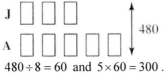

$480 \div 8 = 60$ and $5 \times 60 = 300$.
Ann has 300 beads.

   b. Using proportions, we must set up ratios. Ann has 5 beads for every 3 beads Julie has, so Ann has 5 beads for every 8 beads they have together, or 5:8. let n represent the number of beads Ann has. Together, they have 480 beads.
$\frac{5}{8} = \frac{n}{480}$, $5 \cdot 480 = 8 \cdot n$, $n = \frac{5 \cdot 480}{8}$, $n = 300$.
Ann has 300 beads.

   c. Using algebra, we start with the ratio 3:5 and notice that because of the multiplication property (scaling up), $3 : 5 = 3n : 5n$. Together, they have $3n + 5n$ beads. Then $3n + 5n = 480$, $8n = 480$, $n = 60$. Ann has $5n = 5(60) = 300$ beads.

8.
   b. Using algebra, we can start with the ratios of 5:3 which is equivalent to $5n:3n$, for every counting number n. Then:
$5n - 3n = 4500$
$2n = 4500$, $n = 2250$
$5n = 5 \times 2250 = 11250$, $3n = 3 \times 2250 = 6750$
Smith spent $11,250 and Jones spent $6750.

11. The two are alike because both could represent a ratio. They are different because $\frac{a}{b}$ could represent a number (rational number) or a ratio, but $a:b$ could not represent a number.

13.
   b. 5 ounces

14.
   a. Suppose $a + k : b + k$ is equivalent to $a:b$. Then $\frac{a+k}{b+k} = \frac{a}{b}$, $(a+k)b = (b+k)a$, $ab + kb = ba + ak$, $kb = ak$, $b = a$. In the last step, we can divide by k because k is non-zero. $a + k : b + k$ is equivalent to $a:b$ provided $a = b$.

   b. $ak : bk$ is equivalent to $a:b$ for any nonzero number k (scaling up). So there are no restrictions on a and b.

17. We can demonstrate that there are 35 mice.
   a. Using a ratio table:

# of cats	3	6	9	12	15
# of mice	7	14	21	28	35

   c. Using the unit rate strategy, we start with 3 cats to 7 mice. $3 : 7 = 3 \cdot \frac{1}{3} : 7 \cdot \frac{1}{3} = 1 : \frac{7}{3} = 1 \times 15 : \frac{7}{3} \times 15 = 15 : 35$. There are 35 mice.

20. Since the ratio of apples to oranges is 2:3, we know the ratio of apples to total pieces of fruit is 2:5. This means that $\frac{2}{5}$ of the fruit is apples.

25. The fraction $\frac{a}{b}$ is a number, while the ratio $\frac{a}{b}$ is not a number. In a ratio, the second term b could be 0, but in the fraction $\frac{a}{b}$, b cannot be 0.

27. Numerical values may vary.
   c. The whole-to-part comparison would be there are 7 letters for every 2 As (or there are 7 letters for every 5 Bs).

31.
   a. 108 ounces. The baseline amount would be 90 ounces which is $180 \div 2$. For every 20 minutes of exercise, we need to add 8 ounces of water. Using the unit rate strategy, $20 : 8 = 20 \div 20 : 8 \div 20 = 1 : \frac{8}{20} = 1 : \frac{2}{5}$. So for every 1 minute of exercise, we need an extra $\frac{2}{5}$ ounces of water. Scaling up shows $1 : \frac{2}{5} = 1 \times 45 : \frac{2}{5} \times 45 = 45 : 18$. So we need an additional 18 ounces of water for a total of 108 ounces of water.

36. Using the scaling property,
$2:1:\frac{1}{2} = 2\times 3:1\times 3:\frac{1}{2}\times 3 = 6:3:1\frac{1}{2}$. This shows the totals needed: 6 cups of water, 3 cups of sugar and $1\frac{1}{2}$ cups of lemon juice. He should add 2 more cups of sugar and 1 more cup of lemon juice.

38.
   c. Using the unit rate strategy, we can set up a rate of 2 marbles for Kyle for every 3 more marbles that Marvin has. $2:3 = 2\div 3:3\div 3 = \frac{2}{3}:1 = \frac{2}{3}\times 252:1\times 252 = 168:252$. This shows Kyle has 168 marbles (when Marvin has 252 more marbles than Kyle).

42. Since the two classes have the same ratio, and the girl to boy ratio is 2:3, that means that the girl to total ratio is 2:5. So the number of students must be a multiple of 5. 24 is not a multiple of 5, so there cannot be 24 students in Mrs. Smith's class.

48.
   b. There were 3 boys for every girl.

49. Answers vary. Maria has 3 stamps for every 5 stamps Arnold has, and Maria has 14 fewer stamps than Arnold. How many stamps does each have? (Maria has 21 stamps and Arnold has 35 stamps.)

Chapter 5 Review

2.
   a. The equal sized part is one-eighth.

5.
   a. Using a diagram, we can draw 5 boxes for Erin and 2 for David. Write in the 364 coins together and we have:

   $364 \div 7 = 52$ and $52 \cdot 2 = 104$.
   David has 104 coins.

6.
   b. Using algebra, let e represent the number of coins Erin has. Then David has $\frac{2}{5}e$ coins.
   Then $e - \frac{2}{5}e = 243$, $\frac{5}{5}e - \frac{2}{5}e = 243$,
   $\frac{3}{5}e = 243$, $e = 243 \cdot \frac{5}{3}$, $e = 405$, $\frac{2}{5} \cdot 405 = 162$.
   Erin has 405 coins and David has 162 coins.

9.
   a. The teacher has 9 pencils. She wants to divide them equally among 4 students. How many pencils should each student get?
   b. The teacher has 9 cookies. She wants to divide them equally among 4 students. How much cookie should each student get?

16. The correct answer is tenths, since both fractions are counting the number of tenths.

19. Using algebraic properties:

Equation	Justification
$y + \frac{5}{3} = \frac{9}{4}$	Original expression
$\left(y + \frac{5}{3}\right) + \frac{-5}{3} = \frac{9}{4} + \frac{-5}{3}$	Addition Property of Equality
$y + \left(\frac{5}{3} + \frac{-5}{3}\right) = \frac{9}{4} + \frac{-5}{3}$	Associative Property of Addition
$y + 0 = \frac{9}{4} + \frac{-5}{3}$	Additive Inverse Property
$y = \frac{9}{4} + \frac{-5}{3}$	Addition Identity Property
$y = \frac{27}{12} + \frac{-20}{12}$	Fundamental Law of Fractions
$y = \frac{7}{12}$	Adding Rational Numbers

22. Using the definition of subtraction:

Equation	Justification
$y - \frac{5}{3} = \frac{9}{4}$	Original expression
$y = \frac{9}{4} + \frac{5}{3}$	Definition of subtraction
$y = \frac{27}{12} - \frac{20}{12}$	Fundamental Law of Fractions
$y = \frac{47}{12}$	Adding Rational Numbers
$y = 3\frac{11}{12}$	Simplification

26.
   a. 3/5 of the whole is marked with /'s, and of this region 3/4 of the region is marked with \'s, so this product is 3/4 of 3/5: $\frac{3}{4} \times \frac{3}{5}$.
   b. 2/3 of the whole is marked with /'s, and of this region 2/4 of the region is marked with \'s, so this product is 2/4 of 2/3: $\frac{2}{4} \times \frac{2}{3}$.

29. Using the definition of division:

Equation	Justification
$\frac{3}{5} \times y = 45$	Original expression
$y = 45 \div \frac{3}{5}$	Definition of division
$y = 45 \times \frac{5}{3}$	Using invert and multiply rule
$y = \frac{225}{3}$	Definition of Multiplication of Rational Numbers
$y = 75$	Simplification

30. Using the fact that multiplication and division are inverse operations:

Equation	Justification
$n \div \frac{3}{5} = 28$	Original expression
$n \div \frac{3}{5} \times \frac{3}{5} = 28 \times \frac{3}{5}$	Multiplication Property of Equality
$n = 28 \times \frac{3}{5}$	Multiplication and Division are inverse operations
$n = \frac{84}{5}$	Definition of Multiplication of Rational Numbers
$n = 16\frac{4}{5}$	Simplification

34.
  a. Allia has 4 boxes and Kamran has 3 more compared to Allia, so Kamran has $\frac{3}{4}$ more than Allia.

35.
  b. $\frac{2}{3}$, $1\frac{2}{3}$. Mary has 60 and John has 36, so Mary has $\frac{24}{36} = \frac{2}{3}$ more coins than John, which is $\frac{60}{36} = 1\frac{2}{3}$ times as many coins as John.

36.
  a. $\frac{23}{7} \div \frac{21}{4} = \frac{23}{7} \times \frac{4}{21} = \frac{23 \times 4}{7 \times 21} = \frac{92}{147}$

38. Answers vary.
  a. Samantha frosted $\frac{2}{3}$ of the batch of cookies. Of these cookies, she put sprinkles on $\frac{3}{4}$ of the cookies. What fraction of the batch of cookies had frosting and sprinkles?

43.
  a. Since the diagram has 8 total boxes which represent 72 animals (cats and dogs), then each box must represent $72 \div 8 = 9$ animals. The number of dogs is represented with 5 boxes which would represent $5 \cdot 9 = 45$ dogs.

47.
  a. Using the Make a Table strategy:

Bags	2	4	6	8	10
Water	5	10	15	20	25

  Shehan needs 20 gallons of water.
  c. Using the Unit-rate strategy:
  $2 : 5 = 2 \div 2 : 5 \div 2 = 1 : \frac{5}{2} =$
  $1 \times 8 : \frac{5}{2} \times 8 = 8 : \frac{40}{2} = 8 : 20$.
  So Shehan needs 20 gallons of water.

49. The number of dog boxes is 5 and the number of cat boxes is 3.
  c. $\frac{5}{3} = 1\frac{2}{3}$. Since there are 5 dog boxes for every 3 cat boxes, there are $\frac{5}{3}$ times as many dogs as cats.

50. The number of dog boxes is 9 and the number of cat boxes is 5.
  b. $\frac{4}{9}$. The number of cats is 4 boxes less than the number of dogs. Since there are 9 dog boxes, this decrease would be $\frac{4}{9}$ of the amount of dogs. There are $\frac{4}{9}$ fewer cats than dogs.

Chapter 5 Test

1.
  a. The equal sized part is sixths which comes from the denominator of 6.
  b. There are 8 copies of the equal sized part which comes from the numerator of 8.

4.
  a. To form the diagram, we have 7 boxes for Jim and 3 boxes for Bill. This would be true if Bill had 2 more coins, so we can put 2 fewer coins on Bill. There are 578 total, so put that on the side.

  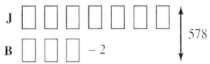

  b. $578 + 2 = 580$, $580 \div 10 = 58$, $3 \cdot 58 - 2 = 174 - 2 = 172$. Bill has 172 coins.

7. We can check the cross products.
  $59 \cdot 11,648 = 687,232$ and $32 \cdot 21,476 = 687,232$. Since the cross products are the same, then $\frac{59}{32} = \frac{21,476}{11,648}$.

9.
  b. $-7\frac{3}{5} = -\left(\frac{7 \times 5 + 3}{5}\right) = -\frac{38}{5}$.

11. 3/5 of the rectangle is marked with /'s, and of this region 2/7 are marked with \'s, so the region marked with × 's represents $\frac{2}{7} \times \frac{3}{5}$.

15. Using the definition of division.

Equation	Justification
$\frac{8}{3} \times y = 60$	Original expression
$y = 60 \div \frac{8}{3}$	Definition of division
$y = 60 \times \frac{3}{8}$	Using invert and multiply rule
$y = \frac{180}{8}$	Definition of Multiplication of Rational Numbers
$y = 22\frac{1}{2}$	Simplification

18. Tanya has 8 boxes and LeBron has 5.
  c. Tanya has 3 boxes more, which is $\frac{3}{5}$ of LeBron's amount. So Tanya has $\frac{3}{5}$ more coins than LeBron.
  d. Tanya has 8 boxes, which is $\frac{8}{5}$ of LeBron's amount. So Tanya has $\frac{8}{5} = 1\frac{3}{5}$ times as many coins as LeBron.

21.
  a. $\dfrac{(3a^{-1})^2 b^4}{12a^{-4}(bc^{-2})^3} = \dfrac{3^2 a^{-2} b^4}{12a^{-4} b^3 c^{-6}} =$
     $\dfrac{9a^4 b^4 c^6}{12a^2 b^3} = \dfrac{3a^2 b c^6}{4}$.
  b. $(2^{-1} b^2)^{-4} = 2^4 b^{-8} = \dfrac{16}{b^8}$.

23. With this strategy, we can see that the ratio of pineapple juice to the total is $\frac{5}{9}$. If we want 60 cups of punch, then the amount of pineapple juice can be represented with p.
$\frac{5}{9} = \frac{p}{60}$, $5 \cdot 60 = 9p$, $300 = 9p$, $\frac{300}{9} = p$,
$p = 33\frac{3}{9} = 33\frac{1}{3}$. So Kerry should add $33\frac{1}{3}$ cups of pineapple juice.

Chapter 6 – Section 6.1

1.
    a. 0.52. There are 4 columns and an additional 12 squares, for a total of $4 \cdot 10 + 12 = 52$ shaded squares out of 100 or a decimal of 0.52.

2.
    a. 0.03 would have 3 shaded squares out of 100.

5. The number 0.42 on a number line is:

    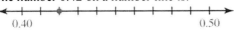

6. In standard form, the numbers are
    a. 347.423

7. In short word form:
    a. 9 and 5 hundredths

8. In word form:
    a. Three and eighty-four thousand, one hundred twenty-three hundred-thousandths

13. Using the square footage of the room in square inches, we would have a room that is $12 \times 12 = 144$ inches by $17 \times 12 = 204$ inches. The room would have an area of $144 \times 204 = 29,376$ square inches. The clay bricks are $3\frac{5}{8}$ inches by $7\frac{5}{8}$ for an area of $3\frac{5}{8} \times 7\frac{5}{8} = 27.640625$ square inches. It would take approximately $29,376 \div 27.640625 \approx 1062.78$ bricks, which is about 1063 bricks. At $0.72 each, this would cost $\$0.72 \times 1063 = \$765.36$. The total cost would be approximately $800 rounded to the nearest hundred.

19. The whole number multiplication required is 348×23.

21.
    b. To the nearest hundredth, the largest interval for 7.23 would be $7.225 \leq n < 7.235$.

22.
    c. 4570.3112; $487.24 \times 9.38 \approx 480 \times 10 = 4800$, so the decimal place location should make the solution close to 4800.

24.
    a. $27,448 - 19,989 = 7459$. So using additive reasoning, we see that a HS graduate earns $7459 more than a HS dropout. Since $27,448 \div 19,989 \approx 1.37$, we could say that a HS graduate earns about 1.37 times as much as a HS dropout.

29.
    a. A square would have an aspect ratio of 1:1, or 3:3, or 9:9. The 4:3 aspect ratio is equivalent to $1.\overline{3}:1$, while the aspect ratio 16:9 is equivalent to $1.\overline{7}:1$. $1.\overline{3}$ is closer to 1, so a TV screen with aspect ratio 4:3 looks more like a square.

32.
    a. Using the ratio, we can create a proportion and solve.
    $\frac{0.27}{0.33} = \frac{n}{90}$, $0.33n = 0.27 \cdot 90$,
    $0.33n = 24.3$, $n \approx 73.64$. We need about 73.6 pounds of water.

36.
    a. The student may be comparing the whole numbers 32 and 7.
    b. If using decimal squares, we can represent the inequality visually. For 0.32, we would shade 3 columns plus 2 more squares. For 0.7, we would shade 7 columns.

    0.32   0.7

    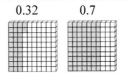

    This illustrates $0.32 < 0.7$.

37.
    a. When we 'bring down the 2', we are really regrouping 1 one and 2 tenths as 12 tenths.

38. e. .23

42. b. 0.2

45. b. 801 and 450 hundred-thousandths.

49. c. 7341.0038 to the nearest thousandth is 7341.004.

53.
    a. We have 6 tenths, and we need to take away 7 tenths. There are not enough tenths, so we regroup 1 one as 10 tenths. There were 3 ones, but one of them was regrouped as 10 tenths. This leaves 2 ones left, represented by crossing out the 3 and writing a 2 above it (3 ones and 6 tenths = 2 ones and 16 tenths).

55. To locate 1.45 on the number line, start by partitioning the number line into tenths. Between 1.2 and 1.7, we will need to mark 1.3, 1.4, …, and 1.6. Then, between 1.4 and 1.5, mark halfway

50

between 1.4 and 1.5 to represent 1.45.

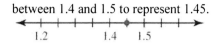

58.
a. Students should first turn this into the whole number multiplication problem 434×26:

$$\begin{array}{r} \overset{2\ 2}{4\,3\,4} \\ \times\ 2\,6 \\ \hline 2604 \\ 8680 \\ \hline 11{,}284 \end{array}$$

The fractional parts of 4.34 and 2.6 have 3 decimal digits together, so the product will also have 3 decimal digits. $4.34 \times 2.6 = 11.284$ is the final answer. This is reasonable because $4.34 \times 2.6 \approx 4 \times 3 = 12$.

65. We can use algebraic equations. Let p represent the number of packs of paper, and m represent the number of markers, so the equations would be $3p + 6m = 4.38$ and $11m + 7p = 9.26$. In the first equation, we could solve for p: $3p + 6m = 4.38$, $3p = 4.38 - 6m$, $p = 1.46 - 2m$. Now we can replace p in the other equation and solve for m. $11m + 7p = 9.26$, $11m + 7(1.46 - 2m) = 9.26$, $11m + 10.22 - 14m = 9.26$, $10.22 - 3m = 9.26$, $-3m = 9.26 - 10.22$, $-3m = -0.96$, $m = 0.32$. Each marker costs $0.32.

69. Answers vary. Other decimal multiplications that would require inserting additional zeros in the product are: $0.2 \times 0.3 = 0.06$, $0.2 \times 0.15 = 0.030$, and $0.1 \times 0.055 = 0.0055$.

Section 6.2

2. Draw a right triangle with legs measuring 2 units and 1 unit. Connect the hypotenuse and the Pythagorean Theorem shows $c^2 = 2^2 + 1^2$, $c^2 = 5$. A negative solution doesn't make sense here, so we can conclude $c = \sqrt{5}$.

4. Answers vary. An equation (without a radical) that would have a solution of $x = \sqrt[3]{25}$ would be $x^3 = 25$ or $4x^3 = 100$. There are some other equations but these are some of the simplest.

7. b. We can let $n = 0.2\overline{182}$. Then $100n = 21.\overline{82}$ and $10{,}000n = 2182.\overline{82}$. Since both of these have the same fractional part, we can subtract them: $10{,}000n - 100n = 2182.\overline{82} - 21.\overline{82}$, $9900n = 2161$, $n = \frac{2161}{9900}$.

11. Since $\frac{1}{333{,}333{,}333} = \frac{3}{999{,}999{,}999}$, we can see that the repetend will be 000 000 003 from the decimal form $\left(0.\overline{000\ 000\ 003}\right)$.

14. a. all   b. some

16. This is an irrational number because 12 is not a perfect square.

18. 9. Since there are 3 non-repeating decimal places, we are really concerned with the 54th digit of the repetend. The repetend has 5 digits, which means it will leave 4 left over after the last full block, which means the 57th digit of the fractional part is 9.

19. Answers vary. An equation that has no radical but does have a solution of $x = \sqrt[5]{-38}$ is $x^5 = -38$.

23. Maria is correct because the product of a non-zero rational number and an irrational number must be irrational. If we let (a, b) be the point on the line, then $b = \sqrt{2} \cdot a$. Since a and b are integers, we can divide and see $\frac{b}{a} = \sqrt{2}$. This would mean $\sqrt{2}$ is a rational number and that cannot happen. So Maria is correct.

25. Answers vary. What we mean is that there are no integers a and b where $\frac{b}{a} = \sqrt{5}$. When we consider the decimal representation, it means that the decimal form of $\sqrt{5}$ is not repeating and not terminating.

27. For every real number x, $x^2 \geq 0$. So the solution to $x^2 = -54$ cannot be a real number.

31. Yes, the student is correct. This number in simplified fraction form would have a denominator with only the prime factors of 2 and 5. This is the definition of terminating decimals.

34. c. Irrational, because 32 is not a perfect square.

36. Answers vary.
a. Some terminating decimals are 4.456, 5.654, 6.4546, etc.
b. A repeating decimal would be $5.\overline{456}$.

c. A non-terminating non-repeating decimal would be something like 4.56566566656666... This type of number is irrational.

39.
   a. With only 2 significant digits, the leading 5 and one more digit would be needed. Rounding appropriately, this would be 5.1×10^{-4}.

41. Answers vary.
   a. $n = 8.\overline{254}$ has a repetend of 254.

43.
   a. $\sqrt{317}$ is irrational because 18 is not a perfect square. It is between 289 and 324, which are consecutive perfect squares; so $\sqrt{317}$ is irrational.

46.
   a. $\sqrt{48} = (48)^{1/2} = (2^4 \times 3)^{1/2} = (2^4)^{1/2}(3)^{1/2} = 2^2\sqrt{3} = 4\sqrt{3}$.

47.
   a. $8x^3 = 12$, $x^3 = \frac{12}{8}$, $x = \sqrt[3]{\frac{12}{8}}$, $x = \frac{\sqrt[3]{12}}{\sqrt[3]{8}}$, $x = \frac{\sqrt[3]{12}}{2}$.

49.
   a. Using a calculator we get $\frac{1514}{999} = 1.\overline{515}$. We could obtain this without a calculator since $\frac{1514}{999} = 1\frac{515}{999} = 1.\overline{515}$.

55. Yes, there could be ordered pairs where both coordinates are irrational and on the line $y = 3x$. For example, if $x = \sqrt{2}$, then $y = 3(\sqrt{2}) = 3\sqrt{2}$. The point $(\sqrt{2}, 3\sqrt{2})$ is on the line $y = 3x$ and both coordinates are irrational.

58.
   a. $\sqrt{167}$ is irrational because 167 is not a perfect square.

61.
   a. $\sqrt{2} + 1$ is an irrational number because it is the sum of an irrational number and a rational number.

62.
   a. Yes. $y^2 + 28 = 8y^2$, $28 = 7y^2$, $y^2 = 4$; there are real numbers that are solutions to this equation.

63.
   b. The principal fourth root of 12 is $\sqrt[4]{12}$.

64. There is only one real solution to $x^7 = 10$ and it is $x = \sqrt[7]{10}$.

67. Proving by contradiction is very effective with these types of problems. So assume that $8\sqrt{3}$ is a rational number, meaning we can write $8\sqrt{3} = \frac{a}{b}$ for some integers a and b. This means $\sqrt{3} = \frac{a}{8b}$ by the division property of equality. $\frac{a}{8b}$ is a rational number because a and $8b$ are integers, which would make $\sqrt{3}$ a rational number. This is impossible so $8\sqrt{3}$ must be irrational.

Section 6.3

1.
   a. Using a bar diagram would look like the following, with first graders on one side and percent on the other. The first vertical bar represents the number of boys and the second vertical bar represents the total.

   First graders n 250
   Percent   48   100

   This creates a proportion which can be solved: $\frac{n}{48} = \frac{250}{100}$, $100n = 48 \cdot 250$, $100n = 12{,}000$, $n = 120$. So 120 first graders are boys.
   b. Using algebra, we can let b represent the number of boys. $b = (48\%)250$, $b = (0.48)250$, $b = 120$. So 120 first graders are boys.

3.
   a. Using a bar diagram would look like the following, with people on one side and percent on the other. The first vertical bar represents the number of people who liked the movie and the second vertical bar represents the total.

   People   45   125
   Percent p 100

   This creates a proportion which can be solved: $\frac{45}{p} = \frac{125}{100}$, $125n = 45 \cdot 100$, $125n = 4500$, $n = 36$. So 36% of the people liked the movie.
   b. Using algebra, we can let p represent the percent of people who liked the movie. $125\left(\frac{p}{100}\right) = 45$, $1.25p = 45$, $p = 36$. Then 36% of the people liked the movie.

7. The percent in decrease from 100% to 64% is 36%.

8. Since he wants to make a profit of $10, then the shirt must be sold (after discount) for $45. The regular price will be higher, so we can call that price R. $R - (15\%)R = 45$, $(85\%)R = 45$, $R = 45 \div 0.85$, $R \approx 52.941176...$ To the nearest penny, the retail price of the shirt is $52.94.

10. If we let the stock purchase price be A, then at the end of the year we have $A - (14\%)A = (86\%)A = B$. If we let p represent the percent increase, then adding the percent increase to the end of the year amount should get us back to the original purchase price: $B + \frac{p}{100}B = A$, $\left(1 + \frac{p}{100}\right)B = A$, $\left(1 + \frac{p}{100}\right) = \frac{A}{B}$, $\left(1 + \frac{p}{100}\right) = \frac{A}{0.86A}$, $\left(1 + \frac{p}{100}\right) = \frac{1}{0.86}$, $1 + \frac{p}{100} = 1 \div 0.86$, $1 + \frac{p}{100} \approx 1.16279$, $\frac{p}{100} \approx 0.16279$, $p \approx 100 \times 0.16279$, $p \approx 16.28\%$. This method illustrates how to find the percentage of increase. Another method is to just plug in the numbers given. $(86\%)A = 0.86A = B$, and combine with $B + 0.1628B = 1.1628B$ to get $1.1628B = 1.1628(0.86A) = 1.000008A$, which is very close to A.

13. The diagram would be nearly identical to the previous problem, but now the girls would be represented with 100 in the percent row as we care comparing boys to girls.

    Children   3   5
    Percent   100 p

    Using a proportion, we can see that the value of p is $\frac{3}{100} = \frac{5}{p}$, $5 \cdot 100 = 3p$, $p = 166\frac{2}{3}$.
    a. Since $166\frac{2}{3}\%$ is more than 100%, there are $166\frac{2}{3}\% - 100\% = 66\frac{2}{3}\%$ more boys than girls. This could be rounded to 66.67%
    b. The number of boys is $166\frac{2}{3}\%$ of the number of girls. This could be rounded to 166.67%.

16. c. 3.5   d. 2.28

19. Since the temperature was 29% higher, it must have been $(129\%)104 = 1.29 \cdot 104 = 134.16$, which is about 134 F.

21.
    a. If we make a bar with customer complaints in one row and percent in the other, we could represent one vertical column with Lexus and the other with Hummer. In this case, there are two diagrams we could make – one with the 100 percent for Lexus and one with 100 percent for Hummer. If the 100 percent was for Hummer, the picture would look like:

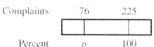

Complaints   76   225
Percent p 100

Then we could set up and solve the proportion $\frac{76}{p} = \frac{225}{100}$, $225p = 76 \cdot 100$, $p = 33\frac{7}{9}$, $p \approx 33.78$. Since 100% is more than 33.78%, we could say that Lexus received about 66.22% fewer complaints. [NOTE: If the bar diagram had 100 for Lexus, we would have the proportion $\frac{76}{100} = \frac{225}{p}$, $76p = 225 \cdot 100$, $p = 296\frac{4}{76}$, $p \approx 296.05$. Since 100% is less than 296.05%, we could say that Hummer received about 196.05% more complaints.]

b. Again, there are two ways to solve this depending on how we represent p, relating to Hummer or Lexus. If we let p represent the percentage of fewer complaints received by Lexus, then the equation would be: $76 = 225 - (p\%)225$, $76 = 225\left(1 - \frac{p}{100}\right)$, $\frac{76}{225} = 1 - \frac{p}{100}$, $1 - \frac{76}{225} = \frac{p}{100}$, $\frac{149}{225} = \frac{p}{100}$, $p = \frac{14900}{225}$, $p \approx 66.22$. Lexus received about 66.22% fewer complaints. [NOTE: If we let p represent the percentage of more complaints received by Hummer, then the equation would be: $225 = 76 + (p\%)76$, $225 = 76\left(1 + \frac{p}{100}\right)$, $\frac{225}{76} = 1 + \frac{p}{100}$, $\frac{225}{76} - 1 = \frac{p}{100}$, $\frac{149}{76} = \frac{p}{100}$, $p = \frac{14900}{76}$, $p \approx 196.05$. Hummer received about 196.05% more complaints.]

23. Answers vary. It could be that inflation is measured as a percentage and the raise reflects raising salaries to match inflation. Or it could be that giving the same raise to all people would mean many people get a significantly higher (or lower) percentage increase. A raise of $2000 is a larger percentage of the salary for a person making $20,000 than for a person making $200,000.

26. If we consider the difference of 26,000 square feet, then this is $\frac{26,000}{62,000} = \frac{26}{62} = \frac{13}{31} \approx 41.94\%$ or $\frac{26,000}{88,000} = \frac{26}{88} = \frac{13}{44} \approx 29.55\%$. But each of these is comparing the difference to a different building. We could correctly say that (1) Building A has about 30% less space than Building B or (2) Building B has about 42% more space than Building A. It looks like he divided correctly but interpreted the result incorrectly. If he meant to describe the percentage more than building A, he should have divided by 88,000 instead of 62,000.

27.
   a. The change in population is 51,000 – 57,600 = –6600 people. The population of San Marcos decreased by 6600 people from 2007 to 2010.
   b. The percent change is $\frac{-6600}{57,600} \cdot 100$
      $= \frac{-6600}{576} = \frac{-1100}{96} = -11.458\overline{3}$, which is about –11.46%. The population in San Marcos decreased by about 11.46% from 2007 to 2010.

29. There are 2 options, so calculate each separately and then compare.
   i. If your initial salary is $36,000 per year, then option A would give a 5% raise on this to end with $(105\%)36,000 = 1.05 \cdot 36,000 = 37,800$.
      The increase for the next year is 12%, ending with $(112\%)37,800 = 1.12 \cdot 37,800 = 42,336$.
      Over the 2 years, this option would give a total of $37,800 + $42,336 = $80,136.
   ii. If your initial salary is $36,000 per year, then option B would give a 12% raise on this to end with $(112\%)36,000 = 1.12 \cdot 36,000 = 40,320$.
      The increase for the next year is 5%, ending with $(105\%)40,320 = 1.05 \cdot 40,320 = 42,336$.
      Over the 2 years, this option would give a total of $40,320 + $42,336 = $82,656.
   iii. If we let the initial salary be p, we can generalize the results. Option A gives $1.05p$ in the first year and $1.176p$ in the second year while Option B gives $1.12p$ in the first year and $1.176p$ in the second year. Both options end with the same amount, but the total paid over the 2 year period is $1.05p + 1.176p = 2.226p$ with Option A compared to $1.12p + 1.176p = 2.296p$ with Option B. Option B is the better choice as it pays more after 1 year, and has a higher total amount after 2 years.

31.
   a. To represent the bar diagram, we can set up calories in one row and percent in the other. The first column would represent lunch while the second column would represent breakfast. Since he consumed 22% fewer calories at lunch, we'll put 100 for the percent at breakfast. Let n represent the total lunch calories. For the percent row, 100 – 22 = 78, so we will put 78 for lunch percent.

   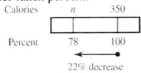

   b. The proportion to solve is
      $\frac{n}{78} = \frac{350}{100}$, $100n = 78 \cdot 350$, $n = 273$.
      Kendall consumed 273 calories at lunch.

35.
   a. To represent the bar diagram, we can set up tuition and fees in one row and percent in the other. The first column would represent 2000/1 while the second column would represent 2010/1. Since there is a 64.4% increase, we'll put 100 for the percent at 2000/1. Let n represent the tuition and fees in 2000/1. For the percent row, 100 + 64.4 = 164.4, so we will put 164.4 for 2010/1 percent.

   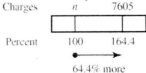

   The proportion to solve is
   $\frac{n}{100} = \frac{7605}{164.4}$, $164.4n = 100 \cdot 7605$,
   $n \approx 4,625.9124...$, $n \approx 4626$. The tuition and fees in 2000/1 were about $4626.
   b. Let n represent the tuition and fees in 2000/1 and since there is a 64.4% increase, we know that $7,605 = n + (64.4\%)n$,
   $7,605 = 1.644n$, $n = \frac{7,605}{1.644}$, $n \approx 4626$. The tuition and fees in 2000/1 were about $4626.

39.
   a. Her new salary is $42,400.
      $40,000 + (6\%)40,000 = (106\%)40,000$
      $= 1.06 \cdot 40,000 = 42,400$.
   b. Her previous salary was about $40,566. If her previous salary was s, the we can solve the equation $s + (6\%)s = 43,000$,
      $(106\%)s = 43,000$, $1.06s = 43,000$,
      $s = \frac{43000}{1.06} \approx 40,566$.

44.
   a. 15%, since 14.67% is closer to 15% than to 14%.
   b. 0.5%, since 0.678% is closer to 0.5% than to 1%.
   c. 134.5%, since 134.53% is closer to 134.5% than to 134.6%.
   d. 140%, since 142.45% is closer to 140% than to 145%.

45.
   a. 10% is one tenth, so we could find 10% of a number by moving the decimal place one place value to the left.

47.
   a. $\frac{17}{20} = \frac{p}{100}$, $20p = 1700$, $p = 85$. So $\frac{17}{20} = 85\%$. Or we can use $\frac{17}{20} = \frac{17 \times 5}{20 \times 5} = \frac{85}{100}$ to find $\frac{17}{20} = 85\%$.

48. b. $320 \times y\% = 400$, $\frac{y}{100} = 400 \div 320$, $y = 125$.

53. From June 2009 to June 2010, the number of complaints increased by $1,419 - 748 = 671$. Comparing this to the 748 starting value, we can create a percentage of increase $\frac{671}{748} \approx 0.8970588... \approx 89.71\%$. This says the number of complaints increased by about 89.71% from June 2009 to June 2010.

57.
   a. Using $A = 25,000$, $n = 60$ months, and $r = 0.063$, we can calculate m:
   $$m = \frac{(25,000) \times \frac{0.063}{12}}{1 - \left(1 + \frac{0.063}{12}\right)^{-60}} = \frac{131.25}{1 - (1.00525)^{-60}},$$
   $m \approx 486.81512...$ So the monthly payment will be rounded up to \$486.82. We can estimate the total payments by multiplying \$468.82 by 60 (the number of payments) and get $\$486.82 \times 60 = \$29,209.20$. Subtracting \$25,000 from this we estimate the interest to be about $\$29,209.20 - \$25,000 = \$4209.20$.

63. \$3.37. Let the regular price be p, so we get the equation: $p - (5\%)p = 3.20$,
   $(95\%)p = 3.20$, $p = \frac{3.20}{0.95}$, $p \approx 3.36842...$ We can round this to the nearest penny as \$3.37.

Chapter 6 Review

2.
   a. 30

4.
   a. 432.045

5. b. $700 + 10 + 6 + \frac{0}{10} + \frac{3}{100} + \frac{4}{1000}$

7. $5 \cdot 10 + 4 \cdot 1 + 3 \cdot 0.1 + 0 \cdot 0.01 + 2 \cdot 0.001$

10.
   a. If a number n is rounded to the nearest hundredth and 2.97 is the result, then $2.965 \leq n < 2.975$.

15. $0.0053 \times 32.7 = (53 \times 10^{-4}) \times (327 \times 10^{-1}) = (53 \times 327) \times (10^{-5})$. So we need to multiply 53×327 and then move the decimal point 5 places to the left.

17.
   a. 132.492, because $27.15 \times 4.88 \approx 30 \times 5 = 150$.

18.
   a. Yes. $\frac{21}{2 \cdot 5 \cdot 7} = \frac{3 \cdot 7}{2 \cdot 5 \cdot 7} = \frac{3}{2 \cdot 5}$ is a terminating decimal because the denominator in simplified form has only prime factors of 2 and 5.

20.
   a. $\frac{34}{99} = 0.\overline{34}$.

21.
   a. Let $n = 12.525252...$, then $100n = 1252.5252...$ and the decimal parts line up exactly. $100n - n = 1252.5\overline{252} - 12.5\overline{252}$
   $99n = 1240$, $n = \frac{1240}{99}$, $n = 12\frac{52}{99}$.

23. $\sqrt{12}$ is irrational because 12 is not a perfect square.

26. c. The solutions to $x^2 = 24$ are $x = \sqrt{24}$ or $x = -\sqrt{24}$.

28.
   a. $\sqrt[3]{4^3 \cdot 5^2 \cdot 8^6} = \sqrt[3]{4^3} \cdot \sqrt[3]{5^2} \cdot \sqrt[3]{8^6} = 4 \cdot \sqrt[3]{5^2} \cdot 8^2 = 256 \cdot \sqrt[3]{25}$.

33.
   b. Since $15\% = 10\% + 5\%$, then we can find 10% of the number and add half of that (representing 5%) to get 15%.

34. Using the bar diagram method:
   a. The diagram would have one row for students and one for percent, with the first vertical bar representing the number of recyclers and the second vertical bar representing the total. We let n represent the number of students who recycle. We put 100 as the total percent and 65 in the recycle percent creating:

   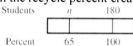

   Using a proportion, we can see that the value of n is $\frac{n}{65} = \frac{180}{100}$, $100n = 65 \cdot 180$, $n = 117$. So 117 students recycle plastic bags.

37.
   b. Using algebra, we can let p represent the percent of people who liked it. $72 = (p\%)400$, $\frac{72}{400} = \frac{p}{100}$, $p = 18$. So 18% of the people in the survey liked the movie.

39.
a. To represent the bar diagram, we can set up pages in one row and percent in the other. The first column would represent the number of pages Pam read while the second column would represent the number of pages Diane read. Since Diane read 65% more pages, we will put 100 in Pam's percent. Let n represent the total pages Pam read. For the percent row, $100 + 65 = 165$, so we will put 165 for Diane's percent.

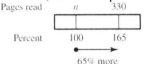

The proportion to solve is
$\frac{n}{100} = \frac{330}{165}$, $165n = 330 \cdot 100$, $n = \frac{33,000}{165}$, $n = 200$. Pam read 200 pages.

43. If the manager makes a 20% profit, then he needs to sell the shirt for $33.60 because
$28 + (20\%)28 = (120\%)28 = 1.20 \times 28 = 33.60$. Let p represent the regular price of the shirt, so we get the equation
$p - (25\%)p = 33.60$, $(75\%)p = 33.60$,
$p = \frac{33.60}{0.75}$, $p = 44.80$. The regular price should be set at $44.80.

Chapter 6 Test

1. b. 0

3. 602.00083

5.
a. $70 + 3 + 0.1 + 0 + 0.005$

6.
a. The simpler problem required is $34,708 \times 142$.
b. The simpler problem required is $34,726.8 \div 4205$.

8.
a. Let $n = 62.35\overline{18}$, then $100n = 6,235.\overline{18}$ and $10,000n = 623,518.\overline{18}$ so the decimal parts line up exactly. Subtracting gives:
$10,000n - 100n = 9,900n$ which is equal to
$623,518.\overline{18} - 6,235.\overline{18} = 617,283$. So
$9,900n = 617,283$, $n = \frac{617,283}{9,900}$, $n = 62\frac{3,483}{9,900}$.

10.
a. Since 15 is not a perfect square, then $\sqrt{15}$ is irrational.

12. a. The real sixths roots of 18 are $\sqrt[6]{18}$ and $-\sqrt[6]{18}$.

18.
a. Using a diagram, we would let the top row represent the number of students, and the bottom row represent percent. The first column would represent 1998 and the second column would represent 2008. Since we know the enrollment increased by 31.7%, then we will put 100 in the percent for 1998 and 100% + 31.7% = 131.7% in the percent for 2008. Let n represent the total number of students in 2008.

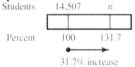

b. The proportion to solve is
$\frac{14,507}{100} = \frac{n}{131.7}$, $100n = 14,507 \cdot 131.7$,
$n = 19,105.719$, $n \approx 19,106$. So there were about 19,106 students in 2008.

21.
a. $0.0046 = \frac{0.0046 \times 100}{100} = \frac{0.46}{100} = 0.46\%$.

23.
a. $312\% = \frac{312}{100} = 3\frac{12}{100} = 3.12$.

Chapter 7 – Section 7.1

2. The function could be written as $f(x) = 3x + 2$, which is 3 times the input ($3x$) and then 2 more than that.

3. b. The table would show inputs and outputs.

x	-2	-1	0	1	2
y	-14	-9	-4	1	6

4.
   a. (3, 11) means that it will cost $11 to mail 3 pounds of clay.
   b. The store charges $14 to mail 4 lb of clay.

5. b. $y(x) = 5x + 3$ coins for Fred where x is the number of coins Amber has.

6. c. $10x + 3y = 6$, $3y = -10x + 6$,
   $y = \frac{-10}{3}x + 2$, $y(x) = -\frac{10}{3}x + 2$

7. c. The function takes the fifths ($5m$) and subtract 1 for the first fifth. Then each of these is assessed a charge of $0.30. So the cab fare in dollars for m miles is $C(m) = 1.60 + 0.30(5m - 1)$, which could be simplified to $C(m) = 1.50m + 1.30$.

11.
    a. Yes, this is a function because for a given finite set, there can be only one output representing the number of elements.
    b. This is not a function because for every non-zero value of B, there are multiple sets possible. That means many inputs have more than one output.

13.
    a. This formula could be correct if the student used x for the vertical sides. Then since 600 feet of wire were used, there would be $600 - 2x$ feet of fence remaining for the top side. The diagram would look like:

    600 – 2x

    x [   ] x

16. a. We know the area is $A = lw$ and $l = 3w + 4$, so we can substitute this into the area function to have only the variable w. $A = lw$, $A(w) = (3w + 4)w$, $A(w) = 3w^2 + 4w$ units².

19. a. x b. k
    c. Based on the function notation, the input is x and the output is $k(x)$ or k.

23.
    a. x represents the location of a term in the sequence: $x = 1$ is the first term, $x = 2$ is the second term, etc.
    b. $f(x)$ represents the actual value of the term in the sequence.

25.
    a. No, this is not a function. There may be many items costing the same amount. For example, there might be two entrees costing $20, so the input of $20 would have 2 outputs.
    b. Yes, if you select an item from the menu, it will have exactly one price.

29.
    a. The domain is {0, 1, 2, 3}, so there will only be 4 points on the graph.

    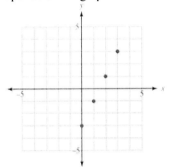

    b. The domain is $0 \leq x \leq 3$, so there will be a line representing all the values of x from 0 to 3.

    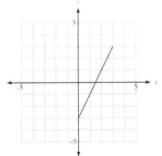

31. a. $j(a) = 4a + 5$

35. a. Since each increase in n by 1 unit increases T by 0.5 units, our equation is $T(n) = 0.5n - 2$.

37.
    a. This expression represents the total number of tiles (black and white) needed for the n-by-($n+1$) patio.
    b. This represents the number of black tiles needed in an n-by-($n+1$) patio. This is obtained by removing 2 units from each dimension of the original patio.

38.
a. If the width of the garden is 24 meters, then the area would be $A(l) = 24l$ m².

42.
a. The speed of the plane is $\frac{600}{3} = 200$ miles per hour.

45.
a. The cost in dollars for one call lasting n minutes is $C(n) = 0.07n + 0.50$.

47.
a. If the boat moves with the current (downstream), then the speed will increase: $d(c) = 48 + c$ miles per hour.

49. It makes sense to define the functions only when they have the same domains. Here, most of the domains match up so we could define $f + g$ on the domain {1, 2, 3, 4}, but it would not make sense to try to define $f + g$ on any other domain because both functions are not defined for other inputs.

51. $f(x) = 4x - 1$ and $g(x) = 2x + 1$.
  a. $(f + g)(-3) = f(-3) + g(-3) = (4(-3) - 1) + (2(-3) + 1) = -18$.
  b. $(f/g)(2) = f(2)/g(2) = \frac{4(2)-1}{2(2)+1} = \frac{7}{5}$.

Section 7.2

1. A solution to the equation is (2.3, – 3.1), which could also be written as $x = 2.3$ and $y = -3.1$.

3.
a. The x-intercepts are where the y-coordinate is 0, so (2,0) and (– 3, 0) are x-intercepts.
b. The y-intercept is where the x-coordinate is 0, so (0, 3) is the y-intercept.
c. To solve $11 = 2 \cdot y(x) + 3$, we can solve for $y(x)$: $11 = 2 \cdot y(x) + 3$, $8 = 2 \cdot y(x)$, $4 = y(x)$.
So we need to find a value of x that has a corresponding y-value of 4. From the table, the point (1, 4) satisfies this condition, so the solution is $x = 1$.

5. a. If we let a represent Alex's age and c represent Chloe's age, then the two equations are $a + c = 29$ and $a = c + 5$.

7.
a. Let m represent the cost in dollars of a bottle of milk and y represent the cost in dollars of a container of yogurt. Then the two equations are $2m + 5y = 18.50$ and $4m + 2y = 17$.

8. b. $40. The daily cost to recycle 13 tons is $y(13) = 40(13) + 130$, $y(13) = 650$, so the cost for the 14th ton would be $y(14) - y(13) = 690 - 650 = 40$. This could also be found with the slope as the cost is going up by $40 per ton.

12. If we let the number of minutes used be represented by m, then plan A will cost $12 + 0.05m$ dollars and plan B will cost $18 + 0.03m$. We can set these two up and see when plan A is less than plan B. $12 + 0.05m < 18 + 0.03m$, $0.02m < 6$, $m < 300$. So after 300 minutes of use, the two plans will be equal. If someone wants to talk more than 300 minutes per month, then I would recommend plan B. If not, then I would recommend plan A.

13.
b. The graph never intersects, so this illustrates that there is no solution to the system of linear equations.

15.
b. The graph illustrates that there are infinitely many solutions because the graphs of the two lines are identical.

21. If we let l represent the length and w represent the width, then the area equation would be $70 = lw$ and $38 = 2l + 2w$. Using substitution, $38 = 2l + 2w$, $2l = 38 - 2w$, $l = 19 - w$, and substituting this into the other equation gives $70 = (19 - w)w$, $70 = 19w - w^2$, $w^2 - 19w + 70 = 0$, $(w - 14)(w - 5) = 0$, $w = 14$ or $w = 5$. Finding the lengths shows the solutions are $w = 14$ and $l = 5$ or $w = 5$ and $l = 14$.

23. Answers vary. The student could graph the equations $y = 3x - 2$ and $y = 4$, finding where they intersect. The student could also graph $y = 3x - 2 - 4 = 3x - 6$ and find when the output is 0 (which would be the x-intercept).

26. The y-intercept occurs when $x = 0$, and since the domain does not include 0, the function cannot have a y-intercept.

29. The student could solve the new equation since it has been created using the substitution method.

33. Let F represent the number of coins Fred has and M represent the number of coins Mark has. Then the equations are $F + M = 200$ and $M = 3F + 8$.

37. Answers vary
   a. 3 x-intercepts and 1 y-intercept:
   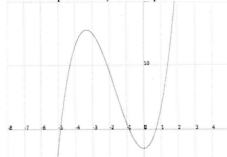

   b. 1 x-intercepts and no y-intercept:
   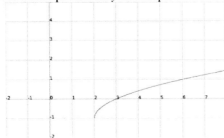

   c. no x-intercepts and 1 y-intercept:
   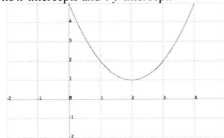

39. $(-2, 4)$ $4x + 12y = 40$, $4x = 40 - 12y$, $x = 10 - 3y$
We can substitute this into the other equation:
$-3(10 - 3y) + 8y = 38$, $-30 + 9y + 8y = 38$,
$17y = 68$, $y = 4$. Then putting this into the first equation we find $x = 10 - 3y$, $x = 10 - 3(4)$, $x = -2$.
Checking the solution:
$-3(-2) + 8(4) = 6 + 32 = 38$ and
$4(-2) + 12(4) = -8 + 48 = 40$. The solution checks!

43. $(4, 2)$. The elimination can be done by adding the equations together and eliminating x.
$$\left.\begin{array}{r}5x + 2y = 24 \\ -5x + 6y = -8\end{array}\right\} \Rightarrow 8y = 16 \Rightarrow y = 2.$$
Substituting this into either equation, we can solve for x. $5x + 2(2) = 24$, $5x = 20$, $x = 4$. Checking the solution: $5(4) + 2(2) = 20 + 4 = 24$ and $-5(4) + 6(2) = -20 + 12 = -8$. The solution checks!

45. $(3, 5)$. The elimination can be done by multiplying the first equation by 4 and then adding the equations together eliminating y.
$$\left.\begin{array}{r}2x - y = 1 \\ 3x + 4y = 29\end{array}\right\} \Rightarrow \left.\begin{array}{r}8x - 4y = 4 \\ 3x + 4y = 29\end{array}\right\} \Rightarrow$$
$11x = 33 \Rightarrow x = 3$.
Substituting this into either equation, we can solve for y. $2(3) - y = 1$, $6 - y = 1$, $y = 5$. Checking the solution: $2(3) - (5) = 6 - 5 = 1$ and $3(3) + 4(5) = 9 + 20 = 29$. The solution checks!

49. Answers vary. Here is one example: Two similar equations are $m = \frac{2}{3}a + 4$ and $m + a = 54$. The word problem would be "Mary has 4 more than 2/3 as many beads as Anne. Together they have 54 beads. How many beads do they each have?" Using substitution, we can find $\left(\frac{2}{3}a + 4\right) + a = 54$, $\frac{5}{3}a = 50$, $a = 30$ which means $m = 24$. So Anne has 30 beads and Mary has 24 beads.

50.
   a. Let d represent the number of dimes she has now and q represent the number of quarters she has now. The equations would be $d + q = 21$ and $10q + 25d = 10d + 25q + 135$.

Section 7.3

1.
   a. $2n + 3 = n + 7$

3.
   a. $2n - 2 = -n - 8$

4. Using tiles:
   a.

5. Using tiles:
   b.

8. b. The conjecture is that $3 \cdot n \div 3 = n$ for all counting numbers n.

10. To start, we will represent the equation:

Then we can add one *n*-tile to both sides.

Simplify the zero pairs.

Then add –4 to both sides.

Simplifying the zero pairs gives:

The solution is $n = -4$.

14. The first mat would involve using substitution to reduce this down to $2(x+1) = x+4$.

17. A multi-step equation involves procedures such as combining like terms, applying the distributive property, and using inverse operations. Examples include $-3x + 2x - 3 = 11$ or $2x - \frac{3}{2}(x - 7) = 8$.

22.
    a. Two-step, because both addition and multiplication must be undone.

25. Start with the equation on the mats.

Then divide each side into two groups.

Lastly, notice that one *n*-tile equals four 1-tiles.

So $n = 4$.

29. Start with the equation on the mats.

$5n + 2 = 3n + 8$
Then subtract two 1-tiles from both sides.

$5n + 2 - 2 = 3n + 8 - 2$
Then simplify.

$5n = 3n + 6$
Then subtract three *n*-tiles from both sides.

$5n - 3n = 3n + 6 - 3n$
Then simplify.

$2n = 6$
Now divide each side into two groups.

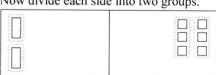

$2n \div 2 = 6 \div 2$
Now simplify for the final solution.

So $n = 3$.

33. Start with the equation: $n - 1 = 7 - n$
On the mats, this looks like:

Then add n to both sides to get:
$n - 1 + n = 7 - n + n$
On the mats, this becomes:

Then simplify the equation to be $2n - 1 = 7$
On the mats, this resembles:

Then add 1 to both sides to get $2n - 1 + 1 = 7 + 1$.
On the mats, this becomes:

Next, simplify to get $2n = 8$
On the mats, this is:

Finally divide both sides by 2 to get $2n \div 2 = 8 \div 2$.
On the mats, this looks like:

Lastly, simplify to get $n = 4$
On the mats, the solution is:

Chapter 7 Review

1.
 a. This is a function because each input maps to exactly one output.

4.
 e. x-coordinate (or first coordinate)
 f. y-coordinate (or second coordinate)

5. c. No, this is not a function because two people could each buy 5 items and spend a different amount of money.

6.
 a. Let k be the number of marbles Kamran has and b represent the number of marbles Bob has. Then $k(b) = b - 5$.

10.
 a. Plugging in the values, we can find k to be $d = kv^2$, $60 = k(50)^2$, $60 = 2500k$, $\frac{60}{2500} = k$, $k = 0.024$.

16.
 b. There are 4 fourths in every mile, so the number of fourths in m miles is $4m$. There will be $4m - 1$ additional fourths because the first one is charged at a higher rate. The function is: $F(m) = 1.45 + 0.30(4m - 1)$.

17.
 b. $T(w) = 420 + w$ is the speed of the plane in miles per hour with a tailwind of w miles per hour.

18.
 c. 15 bags. The 11th foot uses $b(11) - b(10) = 167 - 152 = 15$.

21. The product is 15 but x and y could be either positive or negative. If x decreases, then y must increase.

26. All solutions to the equation $f(x) = 4x + 1$ lie on the graph of the equation.

30. (5, 2) seems like the obvious solution. Checking the point (5, 2) in the equation we get $-4(5) + 9(2) = -20 + 18 = -2$, so the point is a solution.

35. Drawing both on a graph would indicate the solution is $(2, -2)$.

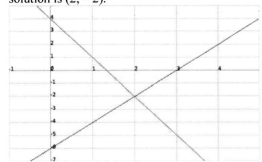

37. $(1, 6)$. Pick an equation and solve for one variable or term that can be substituted into the other equation. Here, pick the first equation and solve for y. $12x - 3y = -6$, $-3y = -12x - 6$, $y = 4x + 2$. Now substitute this into the other equation and solve for x. $-5x + 4y = 19$, $-5x + 4(4x + 2) = 19$, $-5x + 16x + 8 = 19$, $11x = 11$, $x = 1$. Using this in the previous equation shows $y = 4x + 2$, $y = 4(1) + 2$, $y = 6$.

40. $(2, 9)$. Here we can multiply the first equation by 5 and the second by -2 and then add them, which would eliminating the y.
$$\begin{matrix} 2y = -3x + 24 \\ 5y - 2x = 41 \end{matrix} \Rightarrow \begin{matrix} 10y = -15x + 120 \\ 4x - 10y = -82 \end{matrix} \Rightarrow$$
$4x = -15x + 38 \Rightarrow 19x = 38 \Rightarrow x = 2$. Now put this value in any other equation to find the value of y. $2y = -3(2) + 24$, $2y = 18$, $y = 9$.

45. Using tiles to model the equations:
    a.

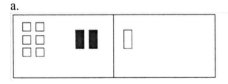

47. To solve the equation with tiles, first set up the equation.

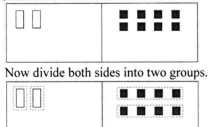

Now divide both sides into two groups.

This simplifies to our final solution.

So $n = -4$.

Chapter 7 Test

2.
   a. This is a function since each x-value corresponds to exactly one y-value.

4.
   a. The independent variable is y.

5.
   a. This equation means that in 4 hours, the company washed 180 windows.

6. d. function

7.
   a. Let w represent the width of a rectangle, then the length of the rectangle, l, is $l(w) = 3w - 5$.

9.
   b. Using function notation with g representing the number of gallons used, $b(g) = 3.50 + 0.003g$ dollars.

10.
   d. $24.60. For an 8 mile ride, $C(8) = 2.60 + 0.40(7(8) - 1)$, $C(8) = 2.60 + 0.40(55)$, $C(8) = 24.60$.

12.
   a. Since the slope is positive, then as Fahrenheit increases, Celsius will also increase.
   b. Using the function, if it is $61.8°$ F, $C = \frac{5}{9}(61.8 - 32)$, $C = 16.55 \approx 16.6$. The temperature is about $16.6°$ C.

14.
   b. From the graph, there are 2 solutions to $-2 = f(x)$. The x-coordinate approximations are -3 and 3.8.

15.
   a. Simplifying first gives us $-1.9 = f(x)$. From the graph, the ordered pair $(-3, -1.9)$ gives the solution $x = -3$.

17. Answers vary, but if $y = 0$, it is easy to find the x-value. $3(0)^2 - 8x = 100$, $-8x = 100$, $x = -12.5$. An ordered pair on the graph is $(-12.5, 0)$.

21. To find the y-intercept, substitute in $x = 0$. This shows $y = 5(0)^2 + 7$, $y = 7$, making $(0, 7)$ the y-intercept. To find the x-intercept(s), substitute in $y = 0$. This shows $0 = 5(x)^2 + 7$, $5x^2 = -7$. This equation has no real number solutions, so there are no x-intercepts on the graph.

24. Using substitution, we can solve the second equation for x: $7y - x = -17$, $7y + 17 = x$. Now substitute this into the other equation to get the value of y:
$4(7y + 17) + 5y = 2$, $28y + 68 + 5y = 2$,
$33y = -66$, $y = -2$. Plugging this into the first equation gives $7(-2) + 17 = x$, $x = 3$, for a solution of $(3, -2)$.

26.
   b. This tells you that the system of equations has infinitely many solutions (the two linear equations represent the same line).
   c. This tells you the system of equations has no solutions (the two linear equations do not intersect; they are parallel).

29. To solve with tiles and algebra, first set up the algebra tiles as the equation.

    This corresponds to the equation
    $3n - 2 = 10 - n$
    Next, we could add one n-tile to both sides.

    This corresponds to the equation
    $3n - 2 + n = 10 - n + n$
    Next, we simplify.

    This corresponds to the equation
    $4n - 2 = 10$
    Next, we add two 1-tiles to both sides.

This corresponds to the equation
$4n - 2 + 2 = 10 + 2$

Next, we simplify.

This corresponds to the equation
$4n = 12$
Next, we divide both sides into four groups.

This corresponds to the equation
$4n \div 4 = 12 \div 4$
Next, we simplify for the final solution.

So $n = 3$.

Chapter 8 – Section 8.1

1.
   a. A tally table is:

Drink preferred	Tally
soda	||
juice	⦀|

   b. A frequency table is:

Drink preferred	Frequency
soda	2
juice	6

   c. A relative frequency table:

Drink preferred	Percent
soda	25%
juice	75%

3. The symbol for 24 beads would be $\frac{24}{32} = \frac{3}{4}$ as tall as the symbol for 32 beads. That would look like:

   32 beads = ▯ , ▯ = 24 beads

5.
   a. Since 4 people out of 9 wanted apples, the angle would be $\frac{4}{9}(360°) = 160°$.

6.
   a. The months which had an increase were April, June, and July.

7.
   c. There are 28 non-majors and 30 majors for 58 total students. Of those, 17 scored in the 70s. This is $\frac{17}{58} = 29\frac{9}{29}\% \approx 29.3\%$.

9.
   b. 3. From the chart, the 5th grade bar is 3 units higher than the 4th grade bar.

10.
    a. numerical (quantitative)
    b. categorical (qualitative)

11. The variables are monthly homeowner fees (numerical), square footage of the house (numerical), zip code (categorical), and type of residence (categorical).

15.
    a. Graph A shows differences between males and females.
    b. Graph B shows the differences among age groups.

17.
    a. The value for grants in 2000-2001 is about 41%.
    b. The value for loans in 1985-1986 is halfway between the 1984-85 year and the 1986-87 year, and is at about 50%.

22.
    b. It could have been that students could select from multiple activities so it was not a "pick only one" survey.

27. Labels for the axes, title, and legend as appropriate.

28. These critical features help the reader interpret the graph properly.

32. c.

33.
    c. To be accurate, we should use the fractional values (non-rounded) to calculate the angles to the nearest degree:

	Calculation	Angle
Yes	$\frac{10}{21} \cdot (360°) = 171\frac{3}{7}°$	171°
No	$\frac{4}{21} \cdot (360°) = 68\frac{4}{7}°$	69°
Undecided	$\frac{7}{21} \cdot (360°) = 120°$	120°

36.
    a. Yes, women with first class accommodations had a higher survival rate.
    b. The bar graph would look like:

    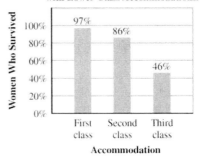

41. A line graph with both men and women would show the trends over time, and an example is:

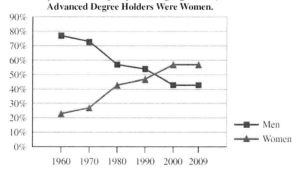

43.
  a. A sample scatterplot is:

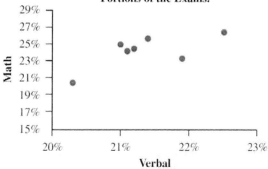

  b. There is a weak positive association between math and verbal scores indicating that as verbal scores increased, so did the math scores.

46.
  a. An appropriate graph would be a line graph because it shows the trend over time and an example is:

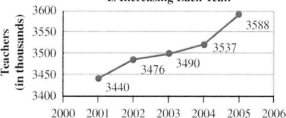

  b. Answers vary. We could predict an increase of about 50 thousand, so 3588 + 50 = 3638, which means about 3,638,000 elementary teachers in 2006.

49.
  a. The years with an increase over previous years were 2001, 2003, 2004, and 2007.

53.
  a. The frequency table is:

interval	frequency
$13 \le x < 16$	2
$16 \le x < 19$	3
$19 \le x < 22$	11
$22 \le x < 25$	5
$25 \le x < 28$	7
$28 \le x < 31$	2

54.
  a. The frequency table is:

interval	frequency
below 50	0
$50 \le x < 55$	1
$55 \le x < 60$	4
$60 \le x < 65$	5
$65 \le x < 70$	7
$70 \le x < 75$	12
$75 \le x < 80$	14
$80 \le x < 85$	5
$85 \le x < 90$	2
above 90	0

  b. A histogram of this data would look like:

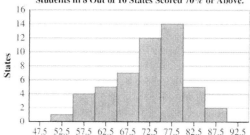

Eighth Graders Who Scored at or Above Basic Proficiency (as a percentage, midpoints of intervals shown)

Section 8.2

1.
  a. The median is more appropriate because of the outlier.

3.
  b. Finding the mean, we end up with an average number of pets: $\bar{x} = \frac{2 \cdot 0 + 5 \cdot 1 + 3 \cdot 2 + 1 \cdot 3}{11} = \frac{14}{11} = 1\frac{3}{11} \approx 1.3$.

5. $\bar{x} = \frac{1 \cdot 2 + 1 \cdot 3 + 3 \cdot 4 + 1 \cdot 5 + 2 \cdot 7 + 4 \cdot 9}{12} = \frac{72}{12} = 6$ for the mean. The median of 12 numbers is found by averaging the 6th and 7th numbers which are 5 and 7, so the median is 6. The mode is 9 because it has the highest frequency.

7. To find the amount per pound, we will need to find the total amount spent and the total number of pounds. With $120 at $15 per pound, there were $\frac{120}{15} = 8$ pounds and with $72 at $18 per pound, there were $\frac{72}{18} = 4$ pounds. The total cost was $120 + $72 = $192 and the total weight was 8 + 4 = 12 pounds, for a per pound average of $\frac{192}{12} = 16$, or $16 per pound.

9.
   a. For odd values of n, the median of n numbers is $x_{\frac{n+1}{2}}$, since the middle term will be in the $\frac{n+1}{2}$ position.
   b. For even numbers, we will average the two terms in the middle which will be in the $\frac{n}{2}$ and $\frac{n}{2}+1$ positions. The formula is $\frac{x_{\frac{n}{2}} + x_{\frac{n}{2}+1}}{2}$.

11.
   a. If the mean for 5 students was 7, then there were a total of 35 sit-ups done. If 4 of the students did only 1 sit-up, then one student could have done 35 − 4 = 31 sit-ups.
   b. If one student did 20 and all students did at least one, then the highest next value would be 35 − 20 − 3 = 12.

14. We can find the mean and median of this set of data. $\bar{x} = \frac{5+3+6+4+5+7+4}{7} = \frac{34}{7} = 4\frac{6}{7} \approx 4.86$, and the median from the ordered data 3, 4, 4, 5, 5, 6, 7 is 5. We could conjecture that about 5 students will miss this question because both the mean and median are close to this value.

15. The interval for usual scores is $\bar{x} \pm 2.5(\text{MAD}) = 72.8 \pm 2.5(4.7) = [61.05, 84.55]$.
   a. Since 65 is in the interval of usual values, it is usual.
   b. Since 87 is not in the interval, it is unusual.

17.
   a. First, the mean is $\bar{x} = \frac{3+5+2+11+3}{5} = \frac{24}{5} = 4.8$. The MAD is then calculated:
   $\text{MAD} = \frac{|4.8-3|+|4.8-5|+|4.8-2|+|4.8-11|+|4.8-3|}{5} \Rightarrow$
   $\text{MAD} = \frac{1.8+0.2+2.8+6.2+1.8}{5} = \frac{12.8}{5} = 2.56 \approx 2.6$.

21.
   a. The mode is the score that occurs most often, which is 80.
   b. There are 30 scores, so the median will be the average of the 15th and 16th values. Both of these are 80, so the median is 80.

   c. The mean can be calculated as:
   $\bar{x} = \frac{50 \cdot 2 + 60 \cdot 3 + 70 \cdot 8 + 80 \cdot 9 + 90 \cdot 5 + 100 \cdot 3}{30} = \frac{2310}{30} = 77$.

25. Rollers. For each, we would calculate the quotient. For the brush, the quotient is $\frac{8-5.30}{2.11} = \frac{2.70}{2.11} \approx 1.28$. For the rollers, the quotient is $\frac{4-2.87}{0.85} = \frac{1.13}{0.85} \approx 1.33$. Since the roller quotient is higher, it is relatively more expensive.

27. Using the average would possibly allow some positive and negative error to cancel, giving a better representation of the overall measurements.

30. One way to interpret this is to put this into a ratio that makes more sense. 1 family to 1.5 pets is a ratio of 1:1.5 = 2:3 = 10:15. So we could interpret this as 3 pets for 2 families or 15 pets for 10 families.

35. If we put all the boys together, we can call their total distance B, and the total distance for the girls G. Then using the mean equations, $\bar{x} = \frac{B}{12}$, $11.2 = \frac{B}{12}$, $B = 134.4$ and $\bar{x} = \frac{G}{15}$, $8.8 = \frac{G}{15}$, $G = 132$. Then the mean for the whole class is $\bar{x} = \frac{B+G}{12+15}$, $\bar{x} = \frac{134.4+132}{27}$, $\bar{x} = \frac{266.4}{27}$, $\bar{x} \approx 9.9$, which is about 9.9 feet on average.

37. d. $\text{MAD} = \frac{|-3.25|+|-4.25|+|4.75|+|2.75|}{4} \Rightarrow$
   $\text{MAD} = \frac{3.25+4.25+4.75+2.75}{4} = \frac{15}{4} = 3.75$.

41. The class with more students would seem to have an unfair advantage, so a fair way should compare the average amount raised. Te second graders raised $\frac{219}{15} = 14.6$ or $14.60 per student, while the third graders raised $\frac{294.40}{23} = 12.8$ or $12.80 per student. When comparing the averages, the second graders did better.

43. Since there are 22 measurements, and the mean is 47 cm, then it must be true that
$\bar{x} = \frac{\text{sum of all measurements}}{22} = 47$ cm. We can solve this for the sum of all measurements to see that the sum of all measurements is $47 \cdot 22 = 1034$ cm. To remove the 36 cm and put in 63 cm, the new sum would be $1034 - 36 + 63 = 1061$ cm. The new mean would be $\bar{x} = \frac{1061}{22} = 48\frac{5}{22} \approx 48.2$ cm. We could accomplish this more quickly by taking the difference $63 - 36 = 27$ and distributing this to all other

values. $\frac{27}{22} = 1\frac{5}{22}$ would be added to the original mean producing $\bar{x} = 47 + 1\frac{5}{22} = 48\frac{5}{22} \approx 48.23$.

46.
   a. Mean, since it can be affected by very high or very low values.
   b. Mean or median, since there will be one and only one value (there could be multiple mode values).

49. Since there are 15 scores and the mean is 78 points, then it must be true that $\bar{x} = \dfrac{\text{sum of all scores}}{15} = 78$. We can solve this for the sum of all scores to see that the sum of all scores is $15 \cdot 78 = 1170$. To add 4 points each for 5 students will add 20 total points, then the sum would be $1170 + 20 = 1190$. The new mean would be $\bar{x} = \frac{1190}{15} = 79\frac{1}{3} \approx 79.33$. We could accomplish this more quickly by noticing that the 20 points would be distributed to all values: $\frac{20}{15} = 1\frac{5}{15} = 1\frac{1}{3}$ would be added to the original mean producing $\bar{x} = 78 + 1\frac{1}{3} = 79\frac{1}{3} \approx 79.33$.

54.
   a. There are 8 data values, so the median will be the average of the 4th and 5th values. The 4th number (in order) is 2.6 and the 5th is 3.4, so the median is $\frac{2.6+3.4}{2} = \frac{6}{2} = 3$. The mean is:
   $\bar{x} = \frac{1.0+2.4+2.6+2.6+3.4+4.0+4.1+4.3}{8} = \frac{24.4}{8} = 3.05$.

56.
   a. The mean for the data is
   $\bar{x} = \frac{26+31+24+40+20+26+29}{7} = \frac{196}{7} = 28$ minutes.
   b. The mean absolute deviation for the data is
   $\text{MAD} = \frac{|26-28|+|31-28|+|24-28|+|40-28|+|20-28|+|26-28|+|29-28|}{7}$
   $\Rightarrow \text{MAD} = \frac{2+3+4+12+8+2+1}{7} = \frac{32}{7} = 4\frac{4}{7} \approx 4.57$ minutes.
   c. The minimum and maximum usual values come from $\bar{x} \pm 2.5(\text{MAD}) = 28 \pm 2.5(4\frac{4}{7})$. The interval of usual values is approximately $[16.6, 39.4]$, so the minimum usual value is about 16.6 minutes and the maximum usual value is about 39.4 minutes.

Section 8.3

1. The 5 number summary is 14, 20, 28, 30, 32.

3.
   a. On average, credit repair is most costly because it has the highest median cost.
   b. The most variation is with credit repair because the IQR (1000 − 100 = 900) is largest of the three types.

6. The IQR for the set of data is $66 - 46 = 20$.
   a. The lowest non-outlier would be at least $Q_1 - 1.5(\text{IQR}) = 46 - 1.5(20) = 46 - 30 = 16$.
   b. The highest non-outlier would be at most $Q_3 + 1.5(\text{IQR}) = 66 + 1.5(20) = 66 + 30 = 96$.

10. Answers vary, but the largest value minus the smallest value must be 25, and $Q_3 - Q_1 = 8$. For example: 20, 30, 35, 38, 45.

12.
   a. $40,850   b. 90%

15. This means 78% of the students who took the test scored less than 85.

16. The whisker length gives a visual sense of the spread of the lower or upper 25% of the data.

22. The box plot would look like

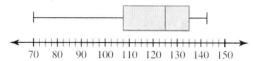

24.
   a. 12 of the 20 prices are lower than 5.3, so it would be the 60th percentile: $\frac{12}{20} = 0.6 = 60\%$.
   d. $\left(\frac{72}{100}\right)20 = 14.4$, which rounds up to 15, and the 15th term is 5.8 cents.

26. The table summary is:

	percentile			number of observations
	25th	50th	75th	
miles	19	22	27	19

28.
   a. 153. There are 64 total terms, so the 50th percentile will be at $\left(\frac{50}{100}\right)64 = 32$. To find the 50th percentile, since this was a whole number, we average the 32nd and 33rd terms: $\frac{150+156}{2} = \frac{306}{2} = 153$.

31. Answers vary, but one situation would be the amount of garbage created – being in the 60th percentile means you produce more garbage than 60% of other families, which is better than producing more than 90% of other families.

33. $Q_1 = 292$ and $Q_3 = 1,789$, so the IQR = 1789 – 292 =1497. Our extreme values would be $Q_1 - 1.5(IQR) = 292 - 1.5(1497) = -1953.5$ and $Q_3 + 1.5(IQR) = 1789 + 1.5(1497) = 4034.5$. Since the values are all positive, there will be no low outliers, but the high outliers will be above 4034.5 meaning salaries (5562) are an outlier.

35.
   a. To start, the low score is 21, with $Q_1 = \frac{44+56}{2} = 50$, $Q_2 = \frac{82+115}{2} = 98.5$, $Q_3 = \frac{125+133}{2} = 129$, and high of 260. The IQR is 129 – 50 = 79, so the outliers would be below $Q_1 - 1.5(IQR) = 50 - 1.5(79) = -68.5$ or above $Q_3 + 1.5(IQR) = 129 + 1.5(79) = 247.5$, meaning 260 is an outlier. The whiskers would be at 21 (low) and 240 (high) with an outlier circle at 260. The boxplot looks like:

   b. To start, the low score is 20, with $Q_1 = \frac{82+112}{2} = 97$, $Q_2 = 120$, $Q_3 = \frac{125+128}{2} = 126.5$, and high of 142. The IQR is 126.5 – 97 = 29.5, so the outliers would be below $Q_1 - 1.5(IQR) = 97 - 1.5(29.5) = 52.75$ or above $Q_3 + 1.5(IQR) = 126.5 + 1.5(29.5) = 170.75$, meaning the values of 20 and 30 are outliers. The whiskers would be at 67 (low) and 142 (high) with outlier circles at 20 and 30. The boxplot looks like:

   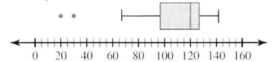

37. To start, the low value is 1,053, with $Q_1 = 1338$, $Q_3 = 2208$, and high of 6015. The IQR is 2208 – 1338 = 870, so the outliers would be below $Q_1 - 1.5(IQR) = 1338 - 1.5(870) = 33$ or above $Q_3 + 1.5(IQR) = 2208 + 1.5(870) = 3513$, meaning only 6015 is an outlier. The whiskers would be at 1053 (low) but there would be no upper whisker because of the outlier at 6015. The boxplot looks like:

   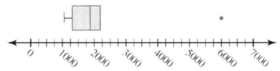

39.
   a. Answers vary, but the outliers appear to be 428 and 445.
   b. $Q_1 = 12$ and $Q_3 = 138$, so the IQR = 138 – 12 = 126. The outliers would be below $Q_1 - 1.5(IQR) = 12 - 1.5(126) = -177$ or above $Q_3 + 1.5(IQR) = 138 + 1.5(126) = 327$. The values 428 and 445 are both outliers.

42.
   a. Because the median for the damage to the car is higher than the median for the SUV, the car (on average) suffers costlier damage.
   b. The car repair cost has more variation because the length of the box (IQR) is larger.

Chapter 8 Review

1.
   a. numerical   b. categorical

2.
   a. The data associated with each participant are reported in the rows.

5.
   a. The frequency table based on manufacturer:

manufacturer	number of recommended products
Samsung	7
Motorola	1
LG	4
Casio	2

9.
   a. The graph suggest that gender affected the survival rate:

   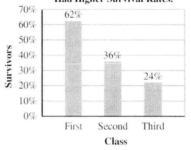

12.
   a. Graph B shows differences in years because the years are the columns.

15.
   a. Graph A exaggerates the small changes.
   b. The vertical axis doesn't start at 0 and the scale is very different from Graph B.

18.
   a. The scatterplot would be:

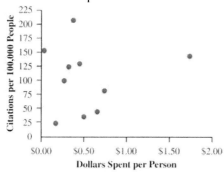

22. The principal can calculate the average amount of money raised (per student) for each group which will take into account the different sizes of the classes.

24. To cancel measurement errors, we can think about using the average measurement (mean), which is $\bar{x} = \frac{14.5+15.1+14.5}{3} = 14.7$ cm.

27. We can find the mean by using the sum of the heights of the boys and the sum of the heights of the girls: $\bar{x} = \frac{14(65)+12(60)}{26} = \frac{1630}{26} = 62\frac{9}{13} \approx 62.7$, so the average height is about 62.7 cm.

31.
   a. The range is $42 - 21 = 21$ feet.
   b. The MAD requires the mean, which is $\bar{x} = \frac{21+24+30+36+38+42}{6} = \frac{191}{6} = 31\frac{5}{6} \approx 31.8$ feet. To be accurate with the MAD, we should use the exact value: $|21-31\frac{5}{6}|+|24-31\frac{5}{6}|+|30-31\frac{5}{6}|+|36-31\frac{5}{6}|+|38-31\frac{5}{6}|+|42-31\frac{5}{6}| = 10\frac{5}{6}+7\frac{5}{6}+1\frac{5}{6}+4\frac{1}{6}+6\frac{1}{6}+10\frac{1}{6} = 41$, so MAD $= \frac{41}{6} = 6\frac{5}{6} \approx 6.8$ or about 6.8 feet.
   c. The minimum and maximum usual values come from $\bar{x} \pm 2.5(\text{MAD}) = 31\frac{5}{6} \pm 2.5(6\frac{5}{6})$ $= 31\frac{5}{6} \pm 17\frac{1}{12}$. The interval of usual values is rounded to the nearest hundredth is $[14.75, 48.92]$, so the minimum usual value is about 14.75 feet and the maximum usual value is about 48.92 feet.

   d. With 6 numbers, the first quartile will be at the $\left(\frac{25}{100}\right) \cdot 6 = 1.5$ term, rounded up this is the 2nd term. The third quartile will be at the $\left(\frac{75}{100}\right) \cdot 6 = 4.5$ term, rounded up this is the 5th term. This means $Q_1 = 24$ and $Q_3 = 38$.

36. We could calculate the quotient of each using the MAD to see who did relatively better. For Samantha, the quotient is $\frac{76-68}{4} = \frac{8}{4} = 2$, and for Mitchell the quotient is $\frac{82-70}{7} = \frac{12}{7} = 1\frac{5}{7} \approx 1.7$. Since $1.7 < 2$, Samantha did relatively better.

37.
   a. The 44th percentile is found with $\left(\frac{44}{100}\right) \cdot 20 = 8.8$ which we round up to be the 9th term, or 43.1 seconds.
   c. There are 8 terms below 43.1, so that accounts for $\frac{8}{20} = 0.40 = 40\%$ of the data. This means 43.1 is the 40th percentile.

40.
   a. In order, the 15 data values are 0, 0, 0, 1, 1, 2, 2, 2, 3, 3, 4, 4, 5, 8, 9. The low value is 0 and the high value is 9, with a median at the 8th term of 2. The other quartiles are at the 4th and 12th term making $Q_1 = 1$ and $Q_3 = 4$. The five number summary is 0, 1, 2, 4, 9.

45.
   a. Since the median for Car B is lower, it has a lower average cost.
   b. The range of values for Car B is greater because of the larger IQR, so there is more variation in repair costs.

47.
   a. Answers vary, but the numbers 19 and 84 appear to be outliers.
   b. $Q_1 = 45$ and $Q_3 = 61$, so the IQR $= 61 - 45 = 16$. The outliers would be below $Q_1 - 1.5(IQR) = 45 - 1.5(16) = 21$ or above $Q_3 + 1.5(IQR) = 61 + 1.5(16) = 85$. The actual outlier is just 19.

Chapter 8 Test

1.
   a. categorical   f. categorical

3.
   a. 7 females between 36 and 50 years old were supporters.

5. The bar graph is:

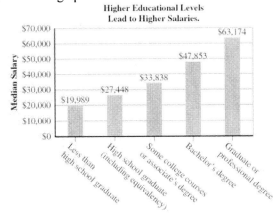

6. The pie chart will need a portion for "other" so that the sum of the values is 100%. The pie chart would look like:

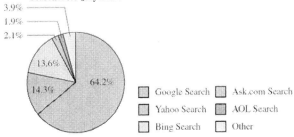

9.
   a. Unemployment rates are highest for those with less than a high school diploma.
   b. The unemployment rises and falls during the cycles of economic expansion and contraction.
   c. Answers vary, but one option is "Lower education levels mean higher unemployment rates."

12.
   a. No apparent association.

16.
   a. x_{201}. With 401 ordered numbers, the 201st number would be the median because there would be 200 numbers above and below this value.
   b. $(x_{412} + x_{413}) \div 2$. Since there are 824 ordered numbers, the median will be the average of the 412th and 413th terms.

18.
   a. The mean is $\bar{x} = \frac{34+38+32+35}{4} = \frac{139}{4} = 34.75$, or about 34.8 miles per gallon.

   b. The MAD is calculated using the exact mean (not the rounded):
   $\text{MAD} = \frac{|34-34.75|+|38-34.75|+|32-34.75|+|35-34.75|}{4} \Rightarrow$
   $\text{MAD} = \frac{0.75+3.25+2.75+0.25}{4} = \frac{7}{4} = 1.75$ or about 1.8 miles per gallon.

19. The minimum and maximum usual values come from $\bar{x} \pm 2.5(\text{MAD}) = 62.8 \pm 2.5(2.4) = 62.8 \pm 6$. The minimum usual value is 56.8 inches and the maximum usual value is 68.8 inches.

21.
   b. There are 29 values below the weight of 74.5 pounds, so the percentile is $\frac{15}{35} = 42\frac{6}{7}\%$, which is about the 83rd percentile.
   c. Since $\left(\frac{70}{100}\right) \cdot 35 = 24.5$, then the 25th term would be used, which is 61.6 pounds.

25.
   a. Team B threw the ball farther on average because the median is higher for team B.
   b. Team B had more variation in distances because the box is wider (which indicates a larger IQR).

Chapter 9 – Section 9.1

NOTE: We will use $P(x)$ to represent the probability of the outcome x. For example, $P(2)$ represents the probability of the outcome 2.

2. This is the event that the arrow of the spinner lands on the number 3, 6, or 7.

3.
   a. The "r" values occur 5 total times, and there are 25 total outcomes. $P(r) = \frac{5}{25} = 20\%$.

4.
   b. The experimental probability from Taylor $\frac{260}{400}$ seems the best estimate because it had a larger value of n.

6.
   a. This represents the probability of randomly drawing a yellow marble.

10. 28. Let n represent the number of blue marbles to add to the bag. We started with 4 blue marbles, so there will be $n + 4$ blue marbles after adding n blue marbles. There are 12 total marbles at the beginning, so adding n blue marbles will give a total of $n + 12$ marbles. We want the probability to be 0.80, so
    $\frac{n+4}{n+12} = 0.80$, $n + 4 = 0.80(n + 12)$,
    $n + 4 = 0.80n + 9.6$, $0.20n = 5.6$, $n = 28$. So we need to add 28 blue marbles to achieve an 80% chance of drawing a blue marble.

13. Let f represent the number of faces with an H. Then $\frac{f}{6} = \frac{1}{3}, f = \left(\frac{1}{3}\right)6, f = 2$. So 2 faces have an H.

15. Let b represent the number of blue marbles, g represent the number of green marbles, and n represent the total number of marbles. Then we know $\frac{b}{n} = 0.25$, $b = 0.25n$ and $\frac{g}{n} = 0.60$, $g = 0.60n$. To find the ratio of blue to green, we can use b:g and then simplify: $b:g = 0.25n : 0.60n = 25:60 = 5:12$.

19.
    a. A table would look like:

# of green	3	6	9	12
# of blue	5	10	15	20

    b. Since there are 3 green marbles for every 5 blue marbles, then there are 3 green marbles for every 8 total marbles in the bag. The probability of drawing a green marble is $\frac{3}{8} = 37.5\%$.

22.
    b. Since $\frac{0.48}{0.15} = 3.2$, we can say that voters are 3.2 times more likely to vote for Proposition A than vote against it. We can also say that registered voters are 3.2 times less likely to vote against Proposition A than for it.

23.
    b. The sum of all outs outcomes is 50, so the probability a player makes an out is $\frac{50}{105} = 47\frac{13}{21}\% \approx 48\%$.

25. This means that if a large number of drawings were done, that about 32% of the time the marble chosen would be green.

29. For this experiment, the researcher would rely on experimental probability because the situation is very complex and is difficult to analyze mathematically. So just tossing the cup and counting would give a good approximation of the theoretical probability.

32.
    a. There are 4 twos and 4 eights, and they do not intersect, so if $E = \{2, 8\}$, $P(E) = \frac{4+4}{52} = \frac{8}{52} = 15\frac{5}{13}\% \approx 15.4\%$.

35. The multiples of 6 are 6, 12, 18, ..., 96, and $96 = 6 \cdot 16$, so there are 16 multiples of 6. This leaves $100 - 16 = 84$ numbers (if 1 and 100 are both included) that are not multiples of 6 so the probability that the number is not a multiple of 6 is $\frac{84}{100} = 84\%$.

36.
    b. There are 6 green marbles, 5 yellow marbles, and 14 total marbles, so the probability of choosing green or yellow is $\frac{6+5}{14} = \frac{11}{14} = 78\frac{4}{7} \approx 78.6\%$.

38. Since 65% is not that close to 50%, we might assume that it was Mark because he only flipped the coin 20 times. The larger number of repetitions for Philip could cause us to assume that his probability is closer to 50%. The short-run results are unpredictable and are more likely to differ from expected results.

40. To design a spinner, we need to find the probability of each piece and then find the central angle.

$P(A) = (3.5)P(B)$. But $P(A) + P(B) = 1$ is also true, so combining gives us $(3.5)P(B) + P(B) = 1$, $(4.5)P(B) = 1$, $P(B) = \frac{1}{4.5}$, $P(B) = \frac{10}{45}$. This means $P(A) = \frac{35}{45}$. The central angles would be $\left(\frac{10}{45}\right) \cdot 360° = 80°$ for B, and $280°$ for A.

42. Each raffle ticket is equally likely, so Bill has the same chance of winning as Ken.

45.
b. If we use the probability to find the actual number of times E occurred (frequency), then we could combine the values together.

Name	P(E)	n	f
Mary	0.58	12	6.96
Bob	0.60	70	42
Tina	0.80	40	32

In Mary's situation, it seems that the probability was rounded to the nearest tenth, so we can use $f = 7$. Then combining gives: $P(E) = \frac{7+42+32}{12+70+40} = \frac{81}{122} = 66\frac{24}{61}\% \approx 66.4\%$.

47.
a. The outcomes with a largest number of 4 come from (4,3), (4,2), (4,1), (1,4), (2,4), (3,4) and (4,4). There are 7 outcomes for this event, with a probability of $\frac{7}{36} \approx 19.4\%$.
b. The outcomes with a largest number of 3 come from (3,2), (3,1), (1,3), (2,3), and (3,3). There are 5 outcomes for this event, with a probability of $\frac{5}{36} \approx 13.9\%$.
c. The event 4 has 7 outcomes, the event 2 has 3 outcomes, and the event 6 has 11 outcomes. The probability of rolling an even number is $\frac{7+3+11}{36} = \frac{21}{36} \approx 58.3\%$.
d. The only numbers less than 3 are 2 and 1, and there are 3 outcomes for the event 2 but only 1 outcome for the event 1. The probability of rolling a number less than 2 is $\frac{3+1}{36} = \frac{4}{36} \approx 11.1\%$.

48. Let n represent the number of packs to purchase. Then $\frac{3}{n} = 0.075$, $3 = 0.075n$, $n = 40$. So we would need to purchase around 40 packs to get 3 coupons.

52. A better interpretation would be that the event E will tend to happen about half of the time. If we repeat the experiment 1000 times, then we would expect the event to occur close to 500 times.

55. Event A is 2.8 times as likely to occur as event B.

57. Event B is 6.2 times as likely to occur as event A.

59. Event B is eight times less likely to occur than event A.

61.
a. To see if A is about 21% less likely than B, we can find the value of
$P(B) - (0.21)P(B) = 0.48 - (0.21)(0.48)$
$= 37.92\%$, which is very close to 38%.
b. To see if B is about 26% more likely than A, we can find the value of
$P(A) + (0.26)P(A) = 0.38 + (0.26)(0.38)$
$= 47.88\%$, which is very close to 48%.
c. To see if B is about 1.26 times as likely as A, we can find the value of
$(1.26)P(A) = (1.26)(0.38) = 47.88\%$, which is very close to 48%.

65. If there is an 18% higher chance, then $p + (18\%)p = p + (0.18)p = (1.18)p$. The probability a person who drinks one glass of juice developing Type II diabetes is $1.18p$.

Section 9.2

1.
a. The Venn diagram is:

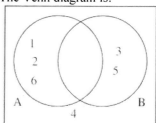

b. The Venn diagram shows that the two events are mutually exclusive because there is no element in the intersection.

3. The expression represents the probability that event A occurs, given that B has occurred.

4.
a. There are 9 dots in set A, and 19 dots in S, so $P(A) = \frac{9}{19}$.

5.
c. There are 3 dots in set "A and B", and 15 dots in S, so $P(A \text{ and } B) = \frac{3}{15} = 20\%$.
e. There are 10 dots in B, and 3 of those dots are in A, so $P(B \mid A) = \frac{3}{10} = 30\%$.

7.
a. Following the branches, we can see the probability of the branch starting with G is $\frac{1}{6}$,

and after that, the probability of Y is $\frac{3}{5}$. So
$P(Y | G) = \frac{3}{5} = 60\%$.

c. Following the branches, we can see the probability of the branch starting with B is $\frac{2}{6}$, and after that, the probability of G is $\frac{1}{5}$. So
$P(BG) = \frac{2}{6} \cdot \frac{1}{5} = \frac{1}{15} = 6\frac{2}{3}\% \approx 6.7\%$.

8.
   a. If we know the first marble is black, then it was placed in box 2 giving four black and one white in that box. The probability of black now is
   $P(\text{Black on 2nd} | \text{Black on 1st}) = \frac{4}{5} = 80\%$.

9. We can draw a tree diagram to illustrate the different branches.

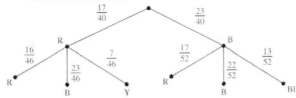

   a. If the first was red, then follow that branch. The probability of black from that point is
   $P(\text{Black on 2nd} | \text{Red on 1st}) = \frac{23}{46} = 50\%$.

10.
   a. The probability is $\frac{1}{50} = 2\%$.
   b. If we know the mother had hypothyroidism, then the probability that the child has IQ lower than 86 is 19%.

13.
   a. Since the events are independent, we can multiply the probabilities:
   $P(A \text{ and } B) = P(A) \cdot P(B) =$
   $(0.42) \cdot (0.75) = 0.315 = 31.5\%$.
   b. $P(A \text{ or } B) = P(A) + P(B) - P(A \text{ and } B)$
   $= 0.42 + 0.75 - 0.315 = 0.855 = 85.5\%$.

15. $P(A \text{ and } B) = P(A) \cdot P(B | A) =$
   $(0.24) \cdot (0.3) = 0.072 = 7.2\%$.

19.
   b. Since each roll of the die is unaffected by previous rolls, the probability of four consecutive rolls that are non-six is
   $\frac{5}{6} \cdot \frac{5}{6} \cdot \frac{5}{6} \cdot \frac{5}{6} = \left(\frac{5}{6}\right)^4 = \frac{625}{1296} = 48\frac{73}{324}\% \approx 48.2\%$.

20.
   b. The probability that there is not a pair of 6s in 24 rolls is $\left(\frac{35}{36}\right)^{24} \approx 50.9\%$.

21. Use C to represent the event that the crime is a property crime and M to represent the event that the crime is a motor vehicle theft.
   b. $P(M | C) = 5.4\%$.

24. If the events were mutually exclusive, then
   $P(A \text{ or } B) = P(A) + P(B)$. Checking this we see
   $P(A \text{ or } B) = 0.6 + 0.7 = 1.3$. This is impossible because the probability of any event, such as event "A or B," cannot exceed 1. So event A and event B are not mutually exclusive.

27. We should not agree. The two events are independent, so the probability that both happen is found by $P(A \text{ and } B) = P(A) \cdot P(B)$. In this instance,
   $P(A \text{ and } B) = \frac{1}{2,300,000} \cdot \frac{1}{2,300,000} \approx 1.89 \times 10^{-13}$. This is extremely unlikely, but still possible.

29.
   a. Yes. We can check this out using
   $P(A \text{ or } B) = P(A) + P(B) - P(A \text{ and } B)$. Putting in our values shows:
   $0.5 = 0.4 + 0.3 - P(A \text{ and } B)$,
   $0.5 = 0.7 - P(A \text{ and } B)$, $0.2 = P(A \text{ and } B)$. So this is possible, as long the intersection of A and B has a probability of 0.2 or 20%.

30. Yes, these events are mutually exclusive because the two events cannot happen at the same time; we can roll a 4 or a 1, but can't roll both on one die.

33. From the dice tables in this chapter, there are 10 ways to get numbers that differ by 1, and 6 ways to get the dice to match. There is no chance these can happen at the same time.
   a. To find the probability that at least one player wins,
   $P(A \text{ or } B) = P(A) + P(B) - P(A \text{ and } B)$,
   $P(A \text{ or } B) = \frac{10}{36} + \frac{6}{36}$, $P(A \text{ or } B) = \frac{16}{36} = \frac{4}{9}$. Then the probability the game ends on the first roll is $\frac{16}{36}$.

37. The draws are independent.
   a. For us to stop after two draws, the first was a non-green and the second was green. There are 7 green marbles and 18 non-green marbles. Since this is with replacement, the total number of marbles remains the same each time.

$P(\text{two draws are needed}) =$
$\frac{18}{25} \cdot \frac{7}{25} = 20\frac{4}{25}\% \approx 20.2\%$.

39. The events are dependent since the number of gumballs (and therefore the probability) change after each gumball is chosen. This is not a replacement situation, so we know she will need to get non-green twice, then a green one.
$P(\text{green on third try}) =$
$\frac{40}{50} \cdot \frac{39}{49} \cdot \frac{10}{48} = 13\frac{13}{49}\% \approx 13.3\%$.

44. In order to end with an odd sum, we must add one even number and one odd number. If spinner A is even, we spin C, and the only odd sum occurs with the outcome A = 2 and C = 3. If spinner A is odd, we spin B and the odd sums occur when B = 4 or B = 2. Finding the probabilities of these requires addition because they are mutually exclusive:
$P(A = 2 \text{ and } C = 3) = (0.8) \cdot (0.4) = 0.32$,
$P(A = 1 \text{ and } B = 2) = (0.2) \cdot (0.5) = 0.1$, and
$P(A = 1 \text{ and } B = 4) = (0.2) \cdot (0.2) = 0.04$. Then
$P(\text{Sum is even}) = 0.32 + 0.1 + 0.04 = 0.46 = 46\%$.

46. There are a total of 100,000 different numbers from 00000 to 99999, and only 10 of those will have all five numbers identical. The correct answer is (b) since Kelly has many more chances to win.

47. We can predict that the proportion of tails (or heads) will be close to 50%. So if we toss the coin 5000 times we should obtain about 2500 tails (or heads).

49.
   a. There are 1573 males who needed follow up care, and 12,165 males who had chronic problems. Let B represent the event that the patient is male, F represent the event that the patient has a follow up visit, and C represent the event that the patient has a chronic problem. Then: $P(B \text{ and } (F \text{ or } C)) =$
   $P(B \text{ and } F) + P(B \text{ and } C) = \frac{1,573}{83,339} + \frac{12,165}{83,339}$
   $= \frac{13,738}{83,339} \approx 16.5\%$.

50.
   b. We could make a table for these outcomes using the probabilities given:

Player A	H	H	H	H	T	T	T	T
Player B	H	H	T	T	H	H	T	T
Player C	H	T	H	T	H	T	H	T
winner		C	B	A	A	B	C	
probability	0.21	0.09	0.14	0.06	0.21	0.09	0.14	0.06

   This shows that each player has 2 chances to win, but they are not equally likely:
   $P(A) = 0.06 + 0.21 = 27\%$,

$P(B) = 0.14 + 0.09 = 23\%$,
$P(C) = 0.14 + 0.09 = 23\%$, and also a 27% chance that no one wins. Player A is favored to win the game.

53.
   a. There were 2871 people who survived out of 3483 total. Let B represent the event that the person survived, then $P(B) = \frac{2871}{3483} \approx 82.4\%$
   b. There are 1150 who were restrained, and of those, 1055 survived. Let A represent the event that the people were restrained and B represent the event that the person survived. Then:
   $P(B | A) = \frac{1055}{1150} \approx 91.7\%$.
   c. If the events were independent, then $P(B | A) = P(B)$. But in part (a) and (b), we see these are not the same. So the events are dependent since $P(B | A) \neq P(B)$.

Section 9.3

1. There are 8 ways A can occur and 12 that $\sim A$ can occur, so the odds that A will occur is $8 : 12 = 2 : 3$.

3. The spinner is broken down so $\frac{115}{360} = \frac{23}{72}$ of the spinner has the value 8 and $\frac{245}{360} = \frac{49}{72}$ of the spinner has the value 17. The expected value would be $e = 8 \cdot \frac{23}{72} + 17 \cdot \frac{49}{72} = \frac{1017}{72} = 14\frac{1}{8} = 14.125$.

5.
   a. The graph from 4U seems to be the least symmetric, so this histogram seems to be a reasonable choice for the summary of only 50 values.
   b. Both L8 and 2B appear fairly symmetric, but perhaps L8 is a bit more symmetric and would fit the normal distribution curve better, so this one would be a good choice for the one with 500 values.

8. A higher z-score means a greater distance from the mean in terms of standard deviations; Mario had a higher z-score and therefore, scored relatively higher on his test.

11. First, we can find Mitchell's z-score: $\frac{86-74}{5} = \frac{12}{5} = 2.4$. In order for Joey to score relatively higher, he needs his z-score to be higher than 2.4. Let x represent Joey's actual score: $\frac{x-72}{10} > 2.4$, $x - 72 > 24$, $x > 96$. If Joey scores above 96, then he will have scored relatively higher than Mitchell.

13. Suppose we obtain 3 heads in a row. Then the probability would be $\frac{1}{2} \cdot \frac{1}{2} \cdot \frac{1}{2} = \frac{1}{8}$, which is the same as the probability for 3 tails in a row. The expected value for this game (without a payment) is $e = (75) \cdot \frac{1}{8} + (45) \cdot \frac{1}{8} + (0) \cdot \frac{6}{8} = 15$. We should expect to win $15 each game on average.

15. C, A, B. Set C has the smallest because the spread of the numbers is the lowest, while set A is next. Set B is the most spread out because of the very low and very high scores.

17. The expected value is
    $e = 0.2 \cdot 3 + 0.15 \cdot 4 + 0.35 \cdot 6 + 0.3 \cdot 7 = 5.4$.

18. We know that about 68% of the data are within one standard deviation of the mean, about 95% of the data are within two standard deviations of the mean and about 99.7% of the data are within three standard deviations. We expect most values to be within 3 standard deviations of the mean. The minimum should be close to $80 - 3 \cdot 5 = 65$, and the maximum close to $80 + 3 \cdot 5 = 95$.

21. If $P(A) = 0.42$, then $P(\sim A) = 0.58$. So the odds in favor are $P(A) : P(\sim A) = 0.42 : 0.58$
    $= 42 : 58 = 21 : 29$.

23. We will find the probability the dart lands in the shaded region and then use it to find the odds.
    $P(\text{dart hits shaded region}) = \frac{\text{Area of shaded region}}{\text{Area of large circle}} =$
    $\frac{\pi(24)^2 - \pi(14)^2}{\pi(24)^2} = \frac{576\pi - 196\pi}{576\pi} = \frac{380\pi}{576\pi} \approx 0.66$. Then the odds in favor of landing in the shaded region are approximately 0.66:34, or 66:34, or 33:17. (This is pretty close to a 2:1 ratio.)

27. There are 5 ways to roll a 6, and 5 ways to roll an 8, so the probability of rolling a 6 or an 8 is $P(A) = \frac{10}{36}$ making $P(\sim A) = \frac{26}{36}$. So the odds of landing on the expensive hotels are $\frac{10}{36} : \frac{26}{36} = 10 : 26 = 5 : 13$, which is very close to a 1:3 ratio.

28.
    b. The odds against prevention are 1:3, so the odds that the state does require prevention are 3:1 making the probability
    $P(A) = \frac{3}{3+1} = \frac{3}{4} = 75\%$.

29. If you take a random sample of many people, ask them how much time they spend on the computer, and then create a histogram for the data, you should expect to see a histogram that is symmetric with the shape of a bell.

33.
   a. Filling out the table, we can put the probability in the corresponding region and use
   $1 - 0.995587 = 0.004413$.

   | Payout | $0 | $250,000 |
   | Probability | 0.995587 | 0.004413 |

   b. The expected value (payout) for each policy is
   $e = 0 \cdot 0.995587 + 250,000 \cdot 0.004413$
   $= 1103.25$, or $1103.25.
   c. To make a profit of $1200 per policy, the company would need to charge $1200 + $1103.25 = $2303.25 per policy.

36. Since the chances are 1 in 13,000, then the probability is $\frac{1}{13,000} \approx 7.7 \times 10^{-5}$.

37.
   a. Turning each odds ratio into probability, the expected value can be estimated as:
   $e \approx \left(\frac{1}{6,800,000}\right) \cdot (1,000) + \left(\frac{1}{6,800,000}\right) \cdot (500)$
   $\left(\frac{1}{6,800,000}\right) \cdot (100) + \left(\frac{97}{70,000}\right) \cdot (25) \approx \$0.035$.
   This is a value of about 3.5 cents per ticket.
   b. A postage stamp costs well more than 3.5 cents, so the expected winnings are less than the cost of the stamp.

39. From the table below, we can subtract the larger from the smaller number to find the number of ways.

−	1	2	3	4	5	6
1	0	1	2	3	4	5
2	1	0	1	2	3	4
3	2	1	0	1	2	3
4	3	2	1	0	1	2
5	4	3	2	1	0	1
6	5	4	3	2	1	0

   There are 6 ways to get a difference of 0, 8 ways to get a difference of 2, and 6 ways to get a difference of 3, so
   $P(\text{losing}) = \frac{6}{36} + \frac{8}{36} + \frac{6}{36} = \frac{20}{36} = \frac{5}{9}$ making
   $P(\text{winning}) = 1 - \frac{5}{9} = \frac{4}{9}$. To break even, the expected value should be 2. Setting up the equation: $e = \frac{5}{9} \cdot 0 + \frac{4}{9} \cdot x$, $2 = 0 + \frac{4x}{9}$,
   $18 = 4x$, $x = 4.5$, making the amount paid out $4.50 to be able to break even.

41.
   a. Since Jack wins 3 out of 7 games, then his probability of winning is approximately
   $P(\text{Jack wins}) = \frac{3}{7}$.

b. Let the digits 0, 1, and 2 represent that Jack won a game, and let the digits 3, 4, 5, and 6 represent that Jill won a game. Each digit from 0 to 6 represents a game played, and we ignore the digits 7, 8, and 9. Each group of three acceptable digits (without any 7's, 8's, or 9's) represents that they played three games. An experiment consists of picking three digits. Repeat this experiment, say, 20 times, and count the number of times f Jack won exactly two games. Based on the table, an estimate of the probability that Jack won exactly two of three games is 7/20 = 35%.

104	210	015	060
223	562	646	205
241	630	304	110
421	514	643	342
365	545	035	365

44. Assume that each prize is equally likely. Let the digits 0, 1, 2, 3, 4, 5, and 6 each represent a different prize. Ignore the digits 7, 8, and 9. An experiment consists of picking a sequence of digits until two identical digits are picked, which represents two identical prizes. Repeat this experiment a large number of times, recording each time the number of boxes k needed to get two identical prizes. Then find the average number of boxes Maria bought. We show the results for 10 experiments: $(4 + 2 + 6 + 7 + 4 + 5 + 4 + 5 + 5 + 3)/10 = 4.5$. Based on these results, on average, Maria can expect to buy 4.5 boxes, on the average, to get two identical prizes.

	# of boxes		# of boxes
1040	4	21060	5
22	2	5622	4
241304	6	63011	5
4216303	7	51434	5
3503	4	545	3

47. We assume the interval [130, 163] represents all measurements that are within 3 standard deviations of the mean. The mean is the center of the interval, so we average the endpoints to determine the center: $\bar{x} \approx (163+130)/2 = 146.5$ cm. All of the data are within 3 standard deviations, so 130 $\approx \bar{x} - 3s$ and $163 \approx \bar{x} + 3s$. Then $163 - 130 \approx 6s$, $33 \approx 6s$, and $s \approx 33/6 = 5.5$. The standard deviation is about 5.5 cm.

49.
a. $\bar{x} = \frac{105+145+150+130}{4} = \frac{530}{4} = 132.5$.

b. The standard deviation is calculated using the table:

Data	Deviation	Squared
105	105 − 132.5 = −27.5	756.25
145	145 − 132.5 = 12.5	156.25
150	150 − 132.5 = 17.5	306.25
130	130 − 132.5 = −2.5	6.25
SUM		1,225

The standard deviation can now be calculated:
$s = \sqrt{\frac{1,225}{4}} = 17.5$.

c. A value of 100 would have a z-score of $\frac{100-132.5}{17.5} = \frac{-32.5}{17.5} = -1\frac{6}{7}$, which is usual since it is less than 2 standard deviations from the mean.

d. A value of 180 would have a z-score of $\frac{180-132.5}{17.5} = \frac{47.5}{17.5} = 2\frac{5}{7}$, which is unusual because it is more than 2 standard deviations from the mean.

51. Drawing a number line to represent the 68-95-99.7 rule, we see:

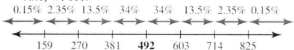

a. The percent of students scoring above 714 is approximately 2.35% + 0.15% = 2.5%.
b. The percent of students scoring less than 381 is approximately 13.5% + 2.35% + 0.15% = 16%.
c. The percentage of scores between 381 and 714 is about 34% + 34% + 13.5% = 81.5%.
d. The percentage of scores between 270 and 492 is about 13.5% + 34% = 47.5%.

Chapter 9 Review

3. 32% means that if you were repeating an experiment 100 times, you could expect event E to occur about 32 times.

4.
b. The sample space is $S = \{1, 2, 3, 4, 5, 6\}$.

6.
b. $P(B) = \frac{15}{22} = 68\frac{2}{11}\% \approx 68.2\%$.

8. An experiment consists of tossing the bent coin three times and noting the outcomes of all three coin tosses (e.g., THT). We repeat this experiment a large number of times (n) and record the number of times (f) all three flips landed on tails. The quotient f/n is an estimate of the probability that all three flips landed on tails. The Law of Large Numbers says that as n increases, then the quotient

f/n approaches the theoretical but unknown probability all three coins land on tails.

10. It seems more likely that Valerie obtained the 35% because she didn't flip the coin very often. Ron's observation should be closer to 50% because of the larger number of repetitions.

13. The number of winning outcomes for player A is $2 + 6 + 5 + 4 = 17$, so player B must have the difference: $36 - 17 = 19$. Since these are equally likely outcomes, player B would win more often. The table of outcomes is found in the chapter, or could be made:

+	1	2	3	4	5	6
1	2	3	4	5	6	7
2	3	4	5	6	7	8
3	4	5	6	7	8	9
4	5	6	7	8	9	10
5	6	7	8	9	10	11
6	7	8	9	10	11	12

16. There are 4 fours and 4 sevens, and they they cannot occur at the same time, so
$P(4 \text{ or } 7) = P(4) + P(7) = \frac{4}{52} + \frac{4}{52} = \frac{8}{52}$
$= 15\frac{5}{13}\% \approx 15.4\%$.

19.
   a. $P(A) \cdot 12 = P(B) \Rightarrow (0.042) \cdot 12 = P(B) \Rightarrow$
      $0.504 = P(B)$.
   b. $P(A) = 5 \cdot P(B) \Rightarrow (0.06) \div 5 = P(B) \Rightarrow$
      $0.012 = P(B)$.
   c. $P(B) = P(A) + (24\%) \cdot P(A) \Rightarrow$
      $P(B) = (0.15) + (0.24)(0.15) \Rightarrow P(B) = 0.186$.
   d. $P(B) = P(A) - (15\%) \cdot P(A) \Rightarrow$
      $P(B) = (0.24) - (0.15)(0.24) \Rightarrow P(B) = 0.204$.

23.
   a. Since there are 4 queens and 52 total cards,
      $P(B) = \frac{4}{52} = 7\frac{9}{13}\% \approx 7.7\%$.
   b. Since a 5 was already drawn, there are still 4 queens but only 51 cards:
      $P(B|A) = \frac{4}{51} = 7\frac{43}{51}\% \approx 7.8\%$.
   c. The events are dependent because $P(B|A) \ne P(B)$.

27.
   c. $P(10 \text{ or ace}) = P(10) + P(\text{ace})$
      $= \frac{4}{52} + \frac{4}{52} = \frac{8}{52} = 15\frac{5}{13}\% \approx 15\%$.

   d. $P(\text{red and face}) = \frac{6}{52} = 11\frac{7}{13}\% \approx 12\%$, or
      $P(\text{red and face}) = P(\text{red}) \cdot P(\text{face}|\text{red}) =$
      $\frac{26}{52} \cdot \frac{6}{26} = \frac{6}{52} = 11\frac{7}{13}\% \approx 12\%$.

28.
   a. There were $11{,}895 + 20{,}446 + 15{,}624 + 10{,}994 + 5336 = 64{,}295$ thousand people from the South out of a total of 182,216 thousand people. The probability is $\frac{64{,}295}{182{,}216} \approx 35\%$

29.
   a. If we let B represent the event that a person has a bachelor's degree and S represent the event that the person is from the South, then
      $P(B|S) = \frac{10{,}994}{64{,}295} \approx 17\%$.

31.
   a. Without replacement, there will be only 51 cards for the second draw:
      $P(2 \text{ first and face card second}) =$
      $P(2 \text{ first}) \cdot P(\text{face card second}|2 \text{ first}) =$
      $\frac{4}{52} \cdot \frac{12}{51} = \frac{12}{663} \approx 1.8\%$.
   b. With replacement, there will be 52 cards for each draw: $P(2 \text{ first and face card second}) =$
      $P(2 \text{ first}) \cdot P(\text{face card second}|2 \text{ first}) =$
      $\frac{4}{52} \cdot \frac{12}{52} = \frac{3}{169} \approx 1.8\%$.

34. There are 10 outcomes in A, 11 outcomes in B and 20 total outcomes.
   a. $P(A) = \frac{10}{20} = 50\%$.
   c. $P(A|B) = \frac{4}{11} = 36\frac{4}{11}\% \approx 36.4\%$.
   f. $P(A \text{ and } B) = \frac{4}{20} = 20\%$.

37. Player B has 11 ways to win, 5 of them from the sum of eight, (1,4), (2,5), (3,6), (4,1), (5,2), and (6, 3). Player A has 12 ways to win: (1,6), (6,1), (2,1), (3,2), (4,3), (5,4), (6,5), (1,2), (2,3), (3,4), (4,5), and (5,6).
   a. $P(A \text{ or } B \text{ wins on first roll}) =$
      $P(A \text{ wins on first roll}) + P(B \text{ wins on first roll})$
      $- P(A \text{ and } B \text{ win on first roll}) =$
      $\frac{12}{36} + \frac{11}{36} = \frac{23}{36} = 63\frac{8}{9}\% \approx 63.9\%$.
   b. The probability that the game continues is $1 - \frac{23}{36} = \frac{13}{36}$ since the probability of a win is $\frac{23}{36}$.
      In order to get to the third roll, the players must have lost twice and then won:
      $P(A \text{ or } B \text{ wins on third roll}) = \frac{13}{36} \cdot \frac{13}{36} \cdot \frac{23}{36} \approx 8.3\%$.

41. Since 8000 people enter the contest, the expected value would be:
$e = \left(\frac{1}{8000}\right) \cdot (2,000) + \left(\frac{2}{8000}\right) \cdot (500) + \left(\frac{3}{8000}\right) \cdot (300)$
$+ \left(\frac{5,994}{8000}\right) \cdot (0) = 0.4875$. This expected value is $0.4875, meaning each ticket on average is worth about 49 cents in prizes.

43.
  a. The minimum number is one, since you could draw yellow on the first draw.
  b. The maximum number is four, since you could draw all the blue and then draw from only the yellow marbles that remain.
  c. The probability of drawing yellow on 1 draw is $\frac{2}{5}$, since there are 2 yellow and 5 total.

  Drawing yellow in 2 draws requires drawing blue first then yellow: $P(BY) = \frac{3}{5} \cdot \frac{2}{4} = \frac{3}{10}$.

  Similarly for 3 draws and 4 draws, $P(BBY) = \frac{3}{5} \cdot \frac{2}{4} \cdot \frac{2}{3} = \frac{1}{5}$ and $P(BBBY) = \frac{3}{5} \cdot \frac{2}{4} \cdot \frac{1}{3} \cdot \frac{2}{2} = \frac{1}{10}$.

outcome	1	2	3	4
probability	$\frac{2}{5}$	$\frac{3}{10}$	$\frac{1}{5}$	$\frac{1}{10}$

  d. We can use the expected value to find the average number of draws:
  $e = \left(\frac{2}{5}\right) \cdot (1) + \left(\frac{3}{10}\right) \cdot (2) + \left(\frac{1}{5}\right) \cdot (3) + \left(\frac{1}{10}\right) \cdot (4) = 2$. It will take an average of 2 draws to get a yellow marble.

46.
  a. $\bar{x} = \frac{126+174+180+156}{4} = \frac{636}{4} = 159$.
  b. The standard deviation is calculated using the table:

Data	Deviation	Squared
126	126 − 159 = − 33	1,089
174	174 − 159 = 15	225
180	180 − 159 = 21	441
156	156 − 159 = − 3	9
SUM		1,764

  The standard deviation can now be calculated: $s = \sqrt{\frac{1,764}{4}} = 21$.

  c. A value of 130 would have a z-score of $\frac{130-159}{21} = \frac{-29}{21} = -1\frac{8}{21}$, which is usual since it is less than 2 standard deviations from the mean.
  d. A value of 205 would have a z-score of $\frac{205-159}{21} = \frac{46}{21} = 2\frac{4}{21}$, which is unusual because it is outside 2 standard deviations of the mean.

50. It may be much easier to draw a number line with standard deviations and obtain approximations from it:

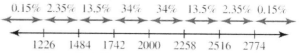

  a. The probability that a randomly chosen ink cartridge prints between 1484 and 2258 pages is 13.5% + 34% + 34% = 81.5%.
  b. The probability that a randomly chosen ink cartridge prints between 1742 and 2774 is 34% + 34% + 13.5% + 2.35% = 83.85%.
  c. The probability that a randomly chosen ink cartridge prints less than 2258 is 2.35% + 13.5% + 34% + 34% = 83.85%.
  d. The probability that a randomly chosen ink cartridge prints more than 2774 is 0.15%.

53. Let the digits 0 and 1 each represent a parent who claimed lack of funding is the biggest problem in school. An experiment consists of picking five random digits, which represent five randomly selected parents. Repeat this experiment a large number n of times, and record the number of times f that you see exactly two of the digits 0 or 1 in the sequence of five digits. The quotient f/n is an estimate of the probability that two out of five randomly selected parents will say lack of funding is the biggest problem at their school. The table shows the results of the simulation. Based on the results, 3/20, or 15%, is an estimate of the probability that two out of five randomly selected parents will say lack of funding is the biggest problem at their school.

10480	15011	**06907**	77921
22368	46573	72905	99562
24130	48360	91977	**96301**
42167	93093	14342	89579
37570	39975	36857	85475

Chapter 9 Test

3.
  a. To see which bag to choose, we can find the probability of drawing blue from each bag.
  $P(\text{blue}) = \frac{8}{13} = 61\frac{7}{13}\% \approx 61.5\%$ for bag A, and $P(\text{blue}) = \frac{45}{75} = 60\%$ for bag B; so it is better to pick from bag A since the probability is higher.
  b. We would expect E to happen about 70 times since $(0.28) \cdot 250 = 70$.

6.
  a. $P(A) = \frac{8}{15} = 53\frac{1}{3}\% \approx 53.3\%$.

b. Since there is no replacement, the number of marbles changes after the first draw:
$P(A|B) = \frac{8}{14} = 57\frac{1}{7}\% \approx 57.1\%$.

c. The events A and B are dependent because $P(A|B) \neq P(A)$.

7.
a. $P(B) = \frac{4}{52} = \frac{1}{13} = 7\frac{9}{13}\% \approx 7.7\%$.

b. Since there is replacement, the number of cards won't change on the second draw.
$P(B|A) = \frac{4}{52} = \frac{1}{13} = 7\frac{9}{13}\% \approx 7.7\%$.

c. These events are independent because $P(B|A) = P(B)$.

10. Without replacement, the values could change and the second marble could be blue two ways: (1) if both are blue and (2) if the first is not blue but the second is. Let A be the event that a blue is drawn first and B be the event that a blue is drawn second. Then, $P(B) = P(A \text{ and } B) + P(\sim A \text{ and } B)$
$= \frac{12}{25} \cdot \frac{11}{24} + \frac{13}{25} \cdot \frac{12}{24} = \frac{288}{600} = 48\%$.

14.
a. For two draws to be needed, a non-green was chosen then a green. Since there is replacement, the total number of marbles will not change from one draw to the next:
$P(\text{non-green first and green second}) =$
$\frac{18}{25} \cdot \frac{7}{25} = \frac{126}{625} = 20.16\% \approx 20.2\%$.

b. For three draws, we need two non-green then one green.
$P(\text{non-green twice and then green third}) =$
$\frac{18}{25} \cdot \frac{18}{25} \cdot \frac{7}{25} = 14.5152\% \approx 14.5\%$.

17. This means if you took a random sample of a large number of women and charted their heights with a histogram, the graph would be symmetric and bell shaped.

19. If necessary, we could create a number line similar to chapter Review #50 or Section 9.3 #51 using the 68-95-99.7 rule.

b. The interval (91.5, 101.5) has a probability of about 34% + 34% + 13.5% + 2.35% = 83.85%.

20. Since there are no outliers, we can expect most of the data to be within 3 standard deviations of the mean. We may assume $28 = \bar{x} - 3s$ and $44 = \bar{x} + 3s$. Then $28 + 44 = 2 \cdot \bar{x}$, and $\bar{x} = (28 + 44)/2 = 36$ grams. (In other words, the mean will be in the middle and we can average the two scores to approximate it.) To find the standard deviation, we note that $28 = \bar{x} - 3s = 36 - 3s$, so $s = (36 - 28)/3 = 8/3 = 2\frac{2}{3}$ grams.

Chapter 10 – Section 10.1

1.
   a. Answers might vary, but three coplanar points are *A*, *H* and *G*. Others on the plane include *C* and *K*.
   c. *G*, *C*, and *K* are collinear points in the plane.

2. If the angles are congruent, then their corresponding measures are equal; statements (b) and (c) are correct.

3. Using mathematical notation:
   c. $\angle ABC \cong \angle DEF$ or $m\angle ABC = m\angle DEF$
   d. $0° < m\angle ABC < 90°$

6.
   a. We draw two coplanar intersecting lines, but could have also drawn non-intersecting coplanar lines (parallel).

   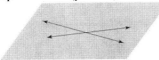

7.
   a.
   b.
   c.
   d.

10. $m\angle A = 3(m\angle B) + 15°$ and $m\angle A + m\angle B = 155°$ are the equations formed. Substituting the value of $m\angle A$ from the first equation gives:
    $3(m\angle B) + 15° + m\angle B = 155° \Rightarrow$
    $4(m\angle B) = 140° \Rightarrow m\angle B = 35°$.
    The sum is $155°$, so $m\angle A = 155° - 35° = 120°$.

11.
    a. 3080. Since $\overleftrightarrow{PQ}$ and $\overleftrightarrow{QP}$ are different names, then we could start by selecting a single point, *P*, and there are 56 ways to choose this point. After that, there are 55 points remaining and we would select one of them to complete the name of the line. This gives $56 \cdot 55 = 3080$ different names.
    b. 1540. Since $\overleftrightarrow{PQ}$ and $\overleftrightarrow{QP}$ represent the same line, then we cannot count them twice. We will have half as many lines as names since each line had 2 names in part (a). $\frac{56 \cdot 55}{2} = 1540$ different lines.

16. Using the angle addition postulate:
    a. $m\angle CPE = m\angle CPD + m\angle DPE \Rightarrow$
       $102° = m\angle CPD + 34° \Rightarrow 68° = m\angle CPD$.
    b. $m\angle APE = m\angle APC + m\angle CPE \Rightarrow$
       $180° = m\angle APC + 102° \Rightarrow 78° = m\angle APC$.
    c. $m\angle APC = m\angle APB + m\angle BPC \Rightarrow$
       $78° = 48° + m\angle BPC \Rightarrow 30° = m\angle BPC$.

18. Choose any point *B* on line *m*, and any point *C* on line *n*. Then the points *A*, *B*, and *C* are three non-collinear points, and by Postulate 2, there will be exactly one plane containing the three points. Since the point *B* and *A* are in the plane, by Postulate 3 the line containing the points (line *m*) lies in the plane. Since the point *C* and *A* are in the plane, by Postulate 3 the line containing the points (line *n*) lies in the plane. Therefore, the plane contains the lines *m* and *n*.

21.
    a. We could start by selecting a single point, and there are 2 ways to choose this point. After that, there is only 1 point remaining and this gives $2 \cdot 1 = 2$ different ways (names).
    b. We could start by selecting a single point, and there are 3 ways to choose this point. After that, there are 2 points remaining and this gives $3 \cdot 2 = 6$ different ways (names).
    c. We could start by selecting a single point, and there are 4 ways to choose this point. After that, there are 3 points remaining and this gives $4 \cdot 3 = 12$ different ways (names).
    d. We could start by selecting a single point, and there are *n* ways to choose this point. After that, there are *n* – 1 points remaining and this gives $n \cdot (n-1)$ different ways (names).

23.
    b.

84

24.
   b.
   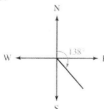

27. Coplanar and noncoplanar describe relationships between points and planes.

31. For both of these names, the order of one letter matters. In $\angle ABC$, the letter B must be in the middle and in $\overrightarrow{DE}$ the D must be first. Also, we could say that each object must have at least one point in every representation of the object. In an angle, the first and last points could change but the vertex point cannot. In a ray, the last point can be any point on the ray, but the first point cannot change. So each has a point that must be included in every possible name of the object.

33.
   a. $\overline{AB}$ or $\overline{BA}$ would work since the order doesn't matter.

35.
   b. Since any 2 points determine a line, we need to have at least 3 points to create a set of noncollinear points.

40.
   a. $\overrightarrow{OK}$ b. HI

43. Since one new line splits 2 original regions into 4 new regions, the number of regions will increase by 2 for every line added.
   a. 3 concurrent lines will divide the plane into $3 \cdot 2 = 6$ regions.

45.
   a.

   Y   Q   B

47. P belongs to the interior of $m\angle ABC$, so $m\angle ABP + m\angle PBC = m\angle ABC$ by the angle addition postulate. Each angle is positive, so $m\angle PBC < m\angle ABP + m\angle PBC = m\angle ABC \Rightarrow m\angle PBC < m\angle ABC$.

55. We need 3 non-collinear points to determine the name for a unique plane. For this plane, there are 32 choices for the first letter, 31 choices for the second, and 30 choices for the last letter for a total of $32 \cdot 31 \cdot 30 = 29,760$ different names.

Section 10.2

1.
   a. No, $ABCD$ is not a valid name because the vertices must be written in a clockwise or counterclockwise order.
   b. Eight total names: $ABCD$, $BCDA$, $CDAB$, $DABC$, $ADCB$, $DCBA$, $CBAD$, and $BADC$.

4. $\angle 1$ and $\angle 2$ are complementary angles.

9.
   a. From the triangle, we see that
   $m\angle ACB + m\angle 1 = 180° \Rightarrow$
   $m\angle 1 = 180° - m\angle ACB$ since they are supplementary. Also, the sum of the angles in a triangle will be $180°$, so
   $m\angle ACB + m\angle CBA + m\angle BAC = 180° \Rightarrow$
   $m\angle CBA + m\angle BAC = 180° - m\angle ACB$.
   Substituting gives
   $m\angle CBA + m\angle BAC = m\angle 1 \Rightarrow$
   $65° + 80° = m\angle 1 \Rightarrow 145° = m\angle 1$.
   b. Using a similar argument as in the previous part, $m\angle CBA + m\angle BAC = m\angle 2 \Rightarrow$
   $40° + 30° = m\angle 2 \Rightarrow 70° = m\angle 2$.

11. $m\angle A + m\angle B + m\angle C = 180°$, so
   $(2n+3)° + n° + (n+5)° = 180° \Rightarrow$
   $4n + 8° = 180° \Rightarrow 4n = 172° \Rightarrow n = 43°$. This means $m\angle B = 43°$, $m\angle C = n + 5° = 48°$, and $m\angle A = 2n + 3° = 89°$.

15. Because it is a triangle, we know
   $m\angle A + m\angle B + m\angle C = 180°$. Since A is the right angle, then $90° + m\angle B + m\angle C = 180° \Rightarrow$
   $m\angle B + m\angle C = 90°$. When two angles sum to $90°$, they are complementary; $\angle B$ and $\angle C$ are complementary.

16.
   a. False, because a trapezoid has exactly one pair of opposite sides parallel, but a square has both pairs of opposite sides parallel.
   b. True, because a rectangle meets the definition of a parallelogram.

18.
   a. Based on the diagram, $\angle 3 \cong \angle 8$ because they are vertical angles. Since $\angle 5 \cong \angle 8$, the transitive property of congruence gives us

∠3 ≅ ∠5. Since the two angles are congruent corresponding angles, we know that $m \parallel n$.

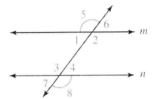

19.

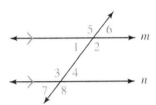

a. Since $m \parallel n$, we know that ∠3 ≅ ∠5 because they are corresponding angles. Because they are vertical angles, ∠3 ≅ ∠8. The transitive property of congruence gives us ∠5 ≅ ∠8.

22.
a. Drawing a picture could help.

Since 5 triangles are created, the sum of the 5 triangles interior angles will be the same as the sum of the interior angles of the heptagon: $5 \cdot 180° = 900°$.

b. Using the sum from part (a), each interior angle would be congruent in a regular heptagon. Then each interior angle would be $\frac{900°}{7} = 128\frac{4}{7} \approx 128.57°$.

c. Interior and exterior angles in a regular heptagon are supplementary. Using the interior angle from part (b) gives each exterior angle a measure of $180° - 128\frac{4}{7}° = 51\frac{3}{7}° \approx 51.43°$.

23.
b. Each interior angle would be congruent in a regular nonagon. Then each interior angle would be $\frac{1260°}{9} = 140°$.

24. To help us find the number of diagonals, we could count the total number of line segments (diagonals and sides) and then remove the sides.

e. Along the same lines as previous parts, an n-gon could be predicted to have $\frac{n(n-1)}{2} - n$ diagonals, which equals $(n^2 - 3n)/2$.

28. A diagonal by definition joins two nonadjacent vertices, but in a triangle every pair of vertices is adjacent.

34. Since ∠A ≅ ∠D and ∠B ≅ ∠E, we know that m∠A = m∠D and m∠B = m∠E. The sum of the interior angles in triangle ABC is m∠A + m∠B + m∠C = 180° and the sum of the interior angles in triangle DEF is m∠D + m∠E + m∠F = 180°. By substitution, m∠A + m∠B + m∠C = m∠D + m∠E + m∠F ⇒ m∠A + m∠B + m∠C = m∠A + m∠B + m∠F ⇒ m∠C = m∠F ⇒ ∠C ≅ ∠F.

36. Using a protractor, we measure 60° for each angle. The conjecture is that the interior angles of an equilateral triangle are each 60°.

39.
a. Based on the diagram, ∠3 ≅ ∠8 and ∠5 ≅ ∠2 because they are vertical angles. Since ∠5 ≅ ∠8, the transitive property of congruence gives us ∠3 ≅ ∠5 and ∠3 ≅ ∠2, making Susan's statement correct.

41.
d. False, since a kite cannot have parallel sides but a trapezoid must have one set of parallel sides.
e. True, since the definition of a square automatically makes it a rectangle.

46. As the problem states, ∠A is an acute angle. m∠A + m∠B + m∠C = 180° ⇒ m∠A + m∠B = 90° and m∠A = 3(m∠B) − 10°. Substituting gives $3(m∠B) - 10° + m∠B = 90° \Rightarrow$ $4(m∠B) = 100° \Rightarrow m∠B = 25°$. Then m∠A = 65°.

49. Since the sum of the interior angles of a quadrilateral is 360°, then m∠A + m∠B + m∠C + m∠D = 360° ⇒ 90° + 90° + 120° + m∠D = 360° ⇒ 300° + m∠D = 360° ⇒ m∠D = 60°.

52.
a. ∠7 and ∠6, ∠4 and ∠6, ∠1 and ∠6.
b. ∠1 and ∠2, ∠4 and ∠2, ∠7 and ∠2, ∠1 and ∠3, ∠4 and ∠3, ∠7 and ∠3, ∠8 and ∠9, ∠9 and ∠5, ∠5 and ∠8.

55.
  a. If there are 32 sides, then each interior angle measures $\frac{30 \cdot 180°}{32} = \frac{5400°}{32} = 168.75°$.
  b. Since interior and exterior angles are supplementary, then the measure of each exterior angle is $180° - 168.75° = 11.25°$.

59. If the sum of the measures of the interior angles is $6,840°$, so we can use the equation $(n-2) \cdot 180° = 6,840°$, and use it to solve for n. $(n-2) \cdot 180° = 6,840° \Rightarrow n-2 = 38 \Rightarrow n = 40$. So this polygon has 40 sides.

Section 10.3

1.
  a. The bases are *ABEG* and *FCDH*.
  b. The bases are *GHIJKL* and *ABCDEF*.

2.
  c. The lateral faces are *CBEF*, *ABED*, and *ACFD*.
  e. The edges are: $\overline{AB}, \overline{AC}, \overline{BC}, \overline{DE}, \overline{DF}, \overline{EF}, \overline{CF}, \overline{AD}, \overline{BE}$.

4.
  a. The bases are regular polygons and are in parallel planes.
  b. The lateral faces are rectangles.

7.
  a. The lateral faces are neither isosceles nor equilateral triangles.
  b. The base is not a regular polygon.

11. Kendra is correct because the name of a regular polyhedron refers to the number of faces. The regular tetrahedron has 4 vertices and 4 faces, so it appears both students are correct. But the regular hexahedron has 6 faces but 8 vertices. So Kendra is correct.

14.
  a. Olivia could be correct because the cone is not a polyhedron.
  b. Leon could be correct because the prism has two bases (or the prism does not have an apex).

20.
  b. $E = 3n$.
  c. Yes. $456 = 3n \Rightarrow n = \frac{456}{3} = 152$. Since n is a whole number, then there will be a prism with 456 edges (the base has 152 sides).
  d. No. $314 = 3n \Rightarrow n = \frac{314}{3} = 104\frac{2}{3}$. Since n is not a whole number, then there will not be a prism with 314 edges.

21.
  b. $E = 2n$.
  c. Yes. $456 = 2n \Rightarrow n = \frac{456}{2} = 228$. Since n is a whole number, then there will be a pyramid with 456 edges (the base has 228 sides).
  d. No. $777 = 2n \Rightarrow n = \frac{777}{2} = 388\frac{1}{2}$. Since n is not a whole number, then there will not be a pyramid with 777 edges.

25.
  a. Each face of a die has the same chance of occurring when regular polyhedrons are used.
  b. Because each face on the die could have a different symbol or outcome, a hexagonal die provides more variety than a tetrahedron.

31. A sphere has no vertices, so it cannot be a polyhedron (or it is not constructed of polygons).

35.
  a. The pyramid has n lateral faces.

36.
  b. The prism has $n + 2$ faces.

37.
  b. A right prism must have lateral faces that are rectangles.

41. A soccer ball is made up of two different regular polygons (hexagons and pentagons), so it cannot be a regular polyhedron. A true soccer ball when inflated would not be a polyhedron because the faces are spherical not polygonal.

45. A pyramid has at most one nontriangular face. *ABCD* and *CDEF* are nontriangular faces; they are quadrilaterals. So the polyhedron cannot be a pyramid.

49.
  a. three b. three c. four d. three e. five

52. (a) pentahedron, because four lateral faces and one base make five total faces and penta is the prefix for five.

Chapter 10 Review

1. a. $\overline{XY} \cong \overline{DE}$

2.
  a. point relationship   e. both

4.
  a. concurrent and coplanar

6. c. reflex

11. $m\angle ABD = 20°$ and $m\angle FBA = 80°$. From the picture, we can see that $m\angle FBD + m\angle DBE + m\angle EBC = 180°$. Putting in the values from the problem,
$100° + 60° + m\angle EBC = 180° \Rightarrow$
$160° + m\angle EBC = 180° \Rightarrow m\angle EBC = 20°$. The angle addition postulate tells us that
$m\angle DBC = m\angle DBE + m\angle EBC$, so
$m\angle DBC = 60° + 20° \Rightarrow m\angle DBC = 80°$. Since $m\angle ABE = m\angle DBC$, we know $m\angle ABE = 80°$ too. By the angle addition postulate,
$m\angle ABE = m\angle ABD + m\angle DBE \Rightarrow$
$80° = m\angle ABD + 60° \Rightarrow 20° = m\angle ABD$. By the angle addition postulate,
$m\angle FBD = m\angle FBA + m\angle ABD \Rightarrow$
$100° = m\angle FBA + 20° \Rightarrow 80° = m\angle FBA$.

12.
    e. complementary angles
    f. alternate exterior angles

15. $m\angle T = 74°$ and $m\angle Q = 16°$. Since $\angle T$ is complementary to $\angle Q$, $m\angle T + m\angle Q = 90°$. We also know that $m\angle T = 5(m\angle Q) - 6°$. Substitution shows:
$m\angle T + m\angle Q = 90° \Rightarrow$
$(5(m\angle Q) - 6°) + m\angle Q = 90° \Rightarrow$
$6(m\angle Q) - 6° = 90° \Rightarrow 6(m\angle Q) = 96° \Rightarrow$
$m\angle Q = 16°$. Using the first equation,
$m\angle T + m\angle Q = 90° \Rightarrow m\angle T + 16° = 90° \Rightarrow$
$m\angle T = 74°$.

18. The upper left 40° angle is vertical with $\angle CAB$, so $m\angle CAB = 40°$. Since $\angle CAB$ is corresponding with $\angle FBE$ and both measure 40°, then the corresponding angles are congruent. By the F Postulate, $\overleftrightarrow{AC} \parallel \overleftrightarrow{DB}$.

22. One interior angle is vertical with 40°, another is supplementary with 75°, while the last one is supplementary with $\angle 1$. So the sum of the interior angles of the triangle is
$40° + 105° + (180° - m\angle 1) = 180°$. We can solve this for the measure of $\angle 1$:
$40° + 105° + (180° - m\angle 1) = 180° \Rightarrow$
$325° - m\angle 1 = 180° \Rightarrow m\angle 1 = 145°$.

25.
    a. If there are 40 sides, then using techniques from previous sections, we can create 38 triangles. Since each triangle has an interior angle sum of 180°, then the sum of the interior angles of a 40 sided polygon is $38(180°) = 6840°$. We could also use the formula $(n-2)180°$ and substitute $n = 40$.

26.
    b. Using the formula $(n-2)180°$, we can substitute $n = 17$ and obtain an interior angle sum of $(17-2)180° = 2700°$.

28. e. exterior angle

31. A polygon is a simple closed curve made up of only line segments (or a simple closed polygonal curve).
    a. It is not a closed curve.

33. a. base

35.
    a. The base is not a regular polygon.

39.
    a. $F = n + 1$, because there are n lateral faces and one base.

41.
    a. A regular polygon is a polygon with n congruent sides and n congruent angles.
    b. There are infinitely many regular polygons, since the number of sides could be any natural number.
    c. A regular polyhedron is a polyhedron such that the faces are exactly one type of regular polygonal and each of the vertices is formed by the intersection of the same number of faces.
    d. There are only 5 regular polyhedron.

43. This shape is a cone.

46. When the edges of the triangle are formed into the pyramid, they will all intersect at the vertex. So point H and point E will end up in the same place, and with both connected to A, we know $AH = AE$.

Chapter 10 Test

1.
   a. There are 5 points, and a line requires two points. So there would be 5 possible first choices and 4 possible second choices, for a total of $5 \cdot 4 = 20$ names.

4.
   a. 24 ways. If we extend $\overline{YZ}$ and create lines that are perpendicular to $\overline{YZ}$ through Y and Z, there will be regions in the upper left, lower left, upper right, and lower right that could be chosen for X. Also, we need to create a circle centered at the midpoint of $\overline{YZ}$ that passes through Y and Z. Then as we shade the regions for obtuse triangles (cannot be on a dashed line), there will be 24 points possible. There are 6 points in the upper left, 6 in the upper right, 4 in the lower left, 4 in the lower right, and 4 inside the circle.

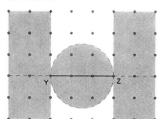

6.
   b. (ii) They are on the same side of the transversal. It is possible that they are congruent, but only if the lines were parallel. It is also possible they are complementary, but not always. The other answer is impossible.

8.
   a. $m\angle DPE = 24°$. From the picture, we can see $m\angle APB + m\angle BPD + m\angle DPE = 180°$. Putting in the values from the problem,
   $46° + 110° + m\angle DPE = 180° \Rightarrow$
   $156° + m\angle DPE = 180° \Rightarrow m\angle DPE = 24°$.

14.
   a. This is a polygon with 16 sides. If it is regular, then we can use $(n-2)180°$ to get the sum of the interior angles: $(16-2)180° = 2520°$. A regular polygon has congruent interior angles, so each one must measure $\frac{(n-2)180°}{n} = \frac{2520°}{16} = 157.5°$. Since interior and exterior angles in polygons are supplementary, the exterior angle must measure $180° - 157.5° = 22.5°$.

16. We know that a regular polygon with n sides will have an interior angle sum of $(n-2)180°$, and that each interior angle will measure $\frac{(n-2)180°}{n}$. Since the angle measures 165°, we can solve
$\frac{(n-2)180°}{n} = 165° \Rightarrow n(180°) - 360° = n(165°) \Rightarrow$
$n(15°) = 360° \Rightarrow n = 24$. This polygon has 24 sides.

24. 30. If there are 58 edges ($E = 58$), then the base must have 29 sides (solve $58 = 2n$ for n), and with 29 sides ($n = 29$) there would be 30 vertices (replace n with 29 in the equation $V = n+1$).

Chapter 11 – Section 11.1

1.
   a. The precision is $\frac{1}{3}$ in because the ruler is marked into thirds.

2.
   d. No; ruler C has higher precision than ruler B, but ruler B provides a more accurate measurement than ruler C. So a higher precision ruler does not always lead to a more accurate measurement. However, in general, you should choose a ruler with higher precision.

5.
   a. kilowatt b. millijoule c. joules per second
   d. nanometer e. second f. megavolt

6.

	Precision	GPE
a. 14.32 lb	0.01 lb	0.005 lb

7.
   a. $\left(\dfrac{5.4 \text{ hats}}{1}\right)\left(\dfrac{4 \text{ flops}}{3 \text{ hats}}\right) = 7.2 \text{ flops}$.

8.
   a. The 4 is the number of units of measurement.
   b. The 'glops' represent the measurement unit.

11. The precision is 0.1 pound, so the GPE is half that or 0.05 pound. $3.2 - 0.05 = 3.15$ and $3.2 + 0.05 = 3.25$, so the true weight is between 3.15 pounds and 3.25 pounds.

12.
   a. $\left(\dfrac{3 \text{ gifs}}{1}\right)\left(\dfrac{7 \text{ lugs}}{13 \text{ gifs}}\right)\left(\dfrac{15 \text{ blips}}{4 \text{ lugs}}\right) = \dfrac{315}{52}$ blips
   $6\frac{3}{52}$ blips ≈ 6.06 blips.
   b. The blip is the smallest unit of measurement. We can find both blips and gifs in terms of lugs to compare. 1 lug $= \frac{15}{4}$ blips $= 3.75$ blips and 1 lug $= \frac{13}{7}$ gifs $= 1\frac{6}{7}$ gifs. So it takes more blips to make 1 lug, meaning that 1 gif is larger than 1 blip, and blips are the smallest.

15. a. 201 in

17. The shortest height would be 5 feet, 8.5 inches and the tallest height would be 5 feet, 9.5 inches.

19.
   a. $\left(\dfrac{340 \text{ ft}}{1}\right)\left(\dfrac{1 \text{ yd}}{3 \text{ ft}}\right) = 113\frac{1}{3}$ yd ≈ 113.33 yd.

23. The length of the nail is $1\frac{1}{6}$ blip because the head of the nail is at $3\frac{2}{6}$ blips while the tip is at $4\frac{3}{6}$ blips, and $4\frac{3}{6} - 3\frac{2}{6} = 1\frac{1}{6}$.

31. We might tell Lilia to use dimensional analysis because it tells what and when to multiply and divide.

36.
   a. To the nearest mile, the segment is 0 mi long.
   b. To the nearest foot, the segment is 0 ft long.
   c. To the nearest inch, the segment is 2 in long.
   d. To the nearest mm, the segment is 62 mm long.

38. a. $\frac{35}{n} = \frac{3}{1} \Rightarrow 3n = 35 \Rightarrow n = \frac{35}{3} = 11\frac{2}{3}$ yd.

40.
   a. With 25 aliens, there would be at total height of $7 \cdot 25 = 175$ yups and $5 \cdot 25 = 125$ peps. Since 12 peps is 1 yup, then we can convert 125 peps into 10 yups and 5 peps. The total height would be 185 yups and 5 peps.

43.
   a. The precision would be 100 m, while the GPE would be 50 m.
   b. The precision would be 1 m, while the GPE would be 0.5 m.

47.
   a. The symbol for day is d.
   b. The symbol for hour is h.
   c. The symbol for minute is min.
   d. The symbol for second is s (or sec).

51.
   a. The grocer should measure the fish from the tip of the nose to the longest part of the tail to get as much length (money) as possible.
   b. The consumer could measure from the tip of the nose to the shortest part of the tail to have less length (money) to pay.

56. a. $2\frac{7}{8}$ units.

58. a. The precision is 0.01 lb, with GPE of 0.005 lb.

Section 11.2

1.
   b. Since the area is 3 snerds, then one snerd would be one-third the size.

3.
   a. To make an obtuse triangle with area of 15 square units, create a base of 6 and height of 5.

   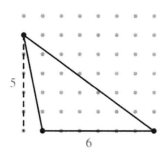

5. This triangle has $AC = 6$, and the height from B perpendicular to $\overleftrightarrow{AC}$ is 5, so the area will be $A = \frac{1}{2}bh = \frac{1}{2} \cdot 6 \cdot 5 = 15$ square units.

7.
   a. The area of the parallelogram is $A = bh = 6 \cdot 3 = 18$ square units.

9.
   a. $4 \text{ km}^2 = 4(10^3 \text{ m})^2 = 4 \cdot 10^6 \text{ m}^2 = 4,000,000 \text{ m}^2$.

11.
   a. $y \times (2y)$ is the area of rectangle $ABCD$.
   b. $1 \times 2y$ is the area of rectangle $EFCD$.

13. A diagram is shown.

	$3a + 4$
$a-3$	$(a-3)(3a+4)$
3	$3(3a+4)$

    with a on the left spanning both rows.

    $a(3a+4) = (a-3)(3a+4) = (a-3) + 3(3a+4)$, so $(a-3)(3a+4) = a(3a+4) - 3(3a+4)$.

15. Let's call w the width and l the length. Then we know that $2w + 2l = 32$ and $wl = 48$. Simplifying, $2w + 2l = 32 \Rightarrow w + l = 16 \Rightarrow w = 16 - l$; now using substitution we see that
    $wl = 48 \Rightarrow (16 - l)l = 48 \Rightarrow 16l - l^2 = 48 \Rightarrow$
    $l^2 - 16l + 48 = 0 \Rightarrow (l-12)(l-4) = 0 \Rightarrow$
    $(l-12) = 0$ or $(l-4) = 0 \Rightarrow l = 12$ or $l = 4$. Solving for w, $w = 16 - l \Rightarrow w = 16 - 12 = 4$
    or $w = 16 - l \Rightarrow w = 16 - 4 = 12$. The dimensions of the rectangle are 12 cm by 4 cm.

16.
   a.

   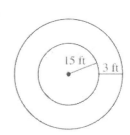

20. We would start by placing one sensor at the end of the fence, and then a sensor every 8 feet. It would take $\frac{425}{8} = 53.125$ or 53 sensors along the fence. We round this down because we need a sensor every 8 feet and the last portion of the fence does not need an additional sensor. There is also the 1 sensor at the beginning for 54 total sensors. Since they come in packs of 3, we will need $\frac{54}{3} = 18$ packs of sensors.

24. We need to find the amount of area wasted, and divide by the total area of the square. The area wasted will be the area of the square minus the area of the circle. Since the radius of the circle is 6 units and the side of the square is 12 units, the amount of area wasted is $12^2 - \pi(6)^2 = 144 - 36\pi$ square units. The area of the square is 144 square units, making the percent wasted $\left(\frac{144-36\pi}{144}\right) \times 100\% = \left(1 - \frac{\pi}{4}\right) \times 100\%$
    $= (100 - 25\pi)\% \approx 21.46\%$.

    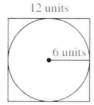

27.
   a. The distance around a circle (circumference) is about 3 times the distance across the circle (diameter) because $C = \pi d$ and $\pi \approx 3$.
   b. The distance around a circle (circumference) is exactly π times the distance across the circle (diameter) because $C = \pi d$.

29.
   a. Using Heron's formula, we need to find the semiperimeter first: $s = \frac{7+12+8}{2} = 13.5$. Then
      $A = \sqrt{s(s-a)(s-b)(s-c)} \Rightarrow$
      $A = \sqrt{13.5(13.5-7)(13.5-12)(13.5-8)} \Rightarrow$
      $A = \sqrt{13.5(6.5)(1.5)(5.5)} \Rightarrow A = \sqrt{723.9375} \Rightarrow$
      $A \approx 26.91 \text{ cm}^2$.

31. Since the circumference is 17 miles, and the proton travels 11,000 revolutions per second, then the proton moves $17 \times 11,000 = 187,000$ miles in one second.

33. Let's call the original height h and base b, and the new height H and new base B. Since we want the area to remain the same, then $\frac{1}{2}bh = \frac{1}{2}BH$. We also know that the base of the triangle is decreased by 20%, so $B = 0.80b$. We can solve for H:
$\frac{1}{2}bh = \frac{1}{2}BH \Rightarrow \frac{1}{2}bh = \frac{1}{2}(0.80b)H \Rightarrow$
$h = 0.8H \Rightarrow \frac{h}{0.80} = H \Rightarrow 1.25h = H$. So the original height must be increased by 25% to keep the area the same.

38.
   a. Let C represent the original circumference and C' represent the new circumference. The original radius is r, and the new radius is $4r$. Then $C = 2\pi \cdot r$ and
   $C' = 2\pi \cdot (4r) \Rightarrow C' = 4(2\pi \cdot r) \Rightarrow C' = 4 \cdot C$,
   meaning that the new circumference is 4 times the original circumference.

   b. Let A represent the original area and A' represent the new area. The original radius is r, and the new radius is $4r$. Then $A = \pi \cdot r^2$ and
   $A' = \pi(4r)^2 \Rightarrow A' = 16(\pi \cdot r^2) \Rightarrow A' = 16 \cdot A$,
   meaning that the new area is 16 times the original area.

42. Answers vary.
   a. To have an area of 12 cm^2, then the product of the height and the sum of the bases must be 24. One way to set this up is to have a height of 3 cm, with one base of 3 cm and another of 5 cm.

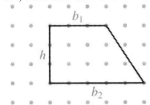

44.
   a. First, we can convert km to cm:
   $\left(\frac{6,378.135 \text{ km}}{1}\right)\left(\frac{10^5 \text{ cm}}{1 \text{ km}}\right) = 637,813,500$ cm.
   Then we can convert cm to miles:
   $\left(\frac{637,813,500 \text{ cm}}{1}\right)\left(\frac{1 \text{ in}}{2.54 \text{ cm}}\right)\left(\frac{1 \text{ ft}}{12 \text{ in}}\right)\left(\frac{1 \text{ mi}}{5280 \text{ ft}}\right)$
   ≈ 3963.189 mi.

48. Using a metric ruler, the base is about 35 mm while the height is about 35 mm giving an area of $A = \frac{1}{2}bh = \frac{1}{2}(35)(35) = 612.5$ mm^2.

51. For the first square, use the variables s_1, P_1, and A_1, and for the second square, use s_2, P_2, and A_2. Then we know $A_1 = s_1^2$ and $A_2 = s_2^2$. Since the areas are the same, we know
$A_1 = A_2 \Rightarrow s_1^2 = s_2^2 \Rightarrow \sqrt{s_1^2} = \sqrt{s_2^2} \Rightarrow s_1 = s_2$. Then $P_1 = 4s_1 = 4s_2 = P_2 \Rightarrow P_1 = P_2$.

55. The radius of circle P could be r, but then the radius of circle Q would be $8r$. The area of circle P is $A_P = \pi \cdot r^2$, and the area of circle Q is
$A_Q = \pi \cdot (8r)^2 = 64(\pi \cdot r^2) = 64 \cdot A_P$. The new area will be 64 times as large as the original area.

56. We can use A for the perimeter of the original rectangle, and A' for the perimeter of the new rectangle. Then $A = \frac{1}{2}bh = \frac{1}{2}(3 \cdot 5) = 7.5$ cm^2, and $A' = \frac{1}{2}bh = \frac{1}{2}(7 \cdot 9) = 31.5$ cm^2.
   a. The area increased by $31.5 - 7.5 = 24$ cm^2.
   b. The area increased by $\frac{24}{7.5} = 3.2 = 320\%$.

61.
   b. To find the area of the triangle using Heron's formula, we need to find the semiperimeter first:
   $s = \frac{7.5+6.1+8.3}{2} = 10.95$. Then
   $A = \sqrt{s(s-a)(s-b)(s-c)} \Rightarrow$
   $A = \sqrt{10.95(10.95-7.5)(10.95-6.1)(10.95-8.3)}$
   $\Rightarrow A = \sqrt{10.95(3.45)(4.85)(2.65)} \Rightarrow$
   $A = \sqrt{485.53531875} \Rightarrow A \approx 22.0$ cm^2.

63.
   a. If the area of a square is 30 m^2, then the sides must be $\sqrt{30}$ m, so the perimeter must be $P = 4\sqrt{30} \approx 21.91$ m. There will be exactly one possibility.

   b. Answers vary. If the area of a rectangle is 30 m^2, then we know $A = lw \Rightarrow 30 = lw \Rightarrow l = \frac{30}{w}$. The perimeter is $P = 2l + 2w \Rightarrow P = 2\left(\frac{30}{w}\right) + 2w$. There are infinitely many values that could be chosen for P. The smallest perimeter will be when the rectangle is a square as in part (a), but there is no largest perimeter. We can conclude that $4\sqrt{30} \leq P$. We can choose any positive value for w and solve $P = 2\left(\frac{30}{w}\right) + 2w$ to find the perimeter.

66. The kite has 2 diagonals, and one diagonal bisects the other as shown.

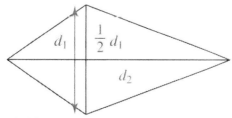

The kite can then be split into two equal sized triangles, each with a height of $\frac{1}{2}d_1$ and a base of d_2, so the area of each triangle is $A = \frac{1}{2}bh = \frac{1}{2}(\frac{1}{2}d_1)d_2 = \frac{1}{4}d_1 \cdot d_2$. The area of the kite is 2 of these triangles: $A = 2(\frac{1}{4}d_1 \cdot d_2) = \frac{1}{2}d_1 \cdot d_2$. So the measurements needed are the lengths of the two diagonals.

69. Since the length is increased by 15%, the new length will be $1.15l$. The new width could be W, and because the areas are the same we know $A = lw$ and $A = (1.15l)W \Rightarrow$
$lw = (1.15l)W \Rightarrow \frac{w}{1.15} = W \Rightarrow W = \frac{20}{23}w$. The percent decrease would be $1 - \frac{20}{23} = \frac{3}{23} = 13\frac{1}{23}\% \approx 13.04\%$. The width needs to decrease by about 13%.

Section 11.3

2.

a.

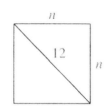

b. Let n represent the length of the side of the square. Then we can solve the Pythagorean theorem to find n: $n^2 + n^2 = 12^2 \Rightarrow$
$2n^2 = 144 \Rightarrow n^2 = 72 \Rightarrow n = \sqrt{72} = 6\sqrt{2}$
≈ 8.49 cm.

4.

a. Solving for $m\angle B$,
$m\angle A + m\angle B + m\angle C = 180° \Rightarrow$
$58° + m\angle B + 43° = 180° \Rightarrow m\angle B = 79°$. $\angle B$ is the largest angle in the triangle, making $\overline{AC}$ the longest side in the triangle.

b. Since $2x + 15$ is greater than x and greater than $x + 5$, $\angle C$ is the largest angle in the triangle making $\overline{AB}$ the longest side in the triangle.

6. We know that $A = lw = 112$ and $l = 4w + 12$, so
$112 = (4w + 12)w \Rightarrow 4w^2 + 12w - 112 = 0$
$\Rightarrow w^2 + 3w - 28 = 0 \Rightarrow (w + 7)(w - 4) = 0 \Rightarrow$
$w = -7$ or $w = 4$. So the width must be 4 m and the length must be $l = 4(4) + 12 = 28$ m. The length of the diagonal is $28^2 + 4^2 = d^2$
$\Rightarrow 800 = d^2 \Rightarrow d = \sqrt{800} = 20\sqrt{2} \approx 28.28$ m.

9.

a. After 30 minutes (one-half of one hour), the legs would measure 26 mi and 20 mi. The distance between them would be found by solving: $26^2 + 20^2 = d^2 \Rightarrow d^2 = 1076 \Rightarrow$
$d = \sqrt{1076} \Rightarrow d \approx 32.80$ mi.

b. After 21 minutes ($\frac{21}{60} = \frac{7}{20}$ of one hour), the legs would measure $\frac{7}{20} \cdot 52 = 18.2$ mi and $\frac{7}{20} \cdot 40 = 14$ mi. The distance between them would be found by solving: $18.2^2 + 14^2 = d^2 \Rightarrow d^2 = 527.24 \Rightarrow$
$d = \sqrt{527.24} \Rightarrow d \approx 22.96$ mi.

11.

a. In order to satisfy the triangle inequality, we need $2 < x < 22$ cm.

b. If $\overline{BC}$ is the shortest side, then the longest side is 12 cm. In order to be acute,
$x^2 + 10^2 > 12^2 \Rightarrow x^2 > 44 \Rightarrow x > \sqrt{44}$. But x is also the shortest side, so we need $\sqrt{44} < x < 10$ cm.

14.

a. It is possible because all sides meet the triangle inequality: $4 + 10 > 13$, $13 + 10 > 4$, and $4 + 13 > 10$.

16.

a. First, find the missing angle:
$m\angle A + m\angle B + m\angle C = 180° \Rightarrow$
$40° + 35° + m\angle C = 180° \Rightarrow m\angle C = 105°$. Since $\angle B$ is the smallest angle, it will be opposite the shortest side $\overline{AC}$.

17.

a. $5^2 + 4^2 = 25 + 16 = 41$ and $8^2 = 64$, so this triangle is obtuse since $5^2 + 4^2 < 8^2$.

18. Answers vary. We can generate our own Pythagorean triple by selecting values for u and v. Let $u = 3$ and $v = 5$, then $a = v^2 - u^2 = 5^2 - 3^2 = 16$, $b = 2uv = 2 \cdot 3 \cdot 5 = 30$, and
$c = v^2 + u^2 = 5^2 + 3^2 = 34$. Checking these values to verify that they form a Pythagorean triple:

$a^2 + b^2 = 16^2 + 30^2 = 1156$ and $c^2 = 1156$, so $a^2 + b^2 = c^2$.

26. Fiona can walk directly north from point A, and Kim can walk directly west from point B. If each person counts the steps until they meet, then the Pythagorean theorem could be used to find the missing length AB.

28. The hypotenuse is the longest side in a right triangle and is opposite the right angle.

29. The Pythagorean equation is $a^2 + b^2 = c^2$.

32. We can let the length be l and the width be w. Since the diagonal is $\sqrt{1664}$ cm, then we know $l^2 + w^2 = \left(\sqrt{1664}\right)^2 \Rightarrow l^2 + w^2 = 1664$. Since the length is 5 times the width, then we know $l = 5w$. Substitution gives us
$(5w)^2 + w^2 = 1664 \Rightarrow 25w^2 + w^2 = 1664 \Rightarrow$
$26w^2 = 1664 \Rightarrow w^2 = 64 \Rightarrow w = \sqrt{64} = 8$ cm. This means $l = 5(8) = 40$ cm.

38. The area of a regular polygon can be found using $A = \frac{1}{2}Pa$. For this situation, $P = 7s$ since this is a heptagon and there are 7 sides. Substitution allows us to solve for s: $A = \frac{1}{2}Pa \Rightarrow 817.6 = \frac{1}{2}(7s)15.6 \Rightarrow$
$817.6 = 54.6s \Rightarrow s = \frac{817.6}{54.6} \approx 15.0$ cm (rounded to nearest tenth)

43.
   a. The area of a regular polygon can be found using $A = \frac{1}{2}Pa$. For this situation, $A \approx \frac{1}{2}(60)(8.7) \approx 261 \, \text{cm}^2$.
   b. The perimeter is 60, so each side is $\frac{60}{6} = 10$ cm. We can find the radius with the Pythagorean theorem: $r^2 = a^2 + \left(\frac{s}{2}\right)^2 \Rightarrow$
   $r^2 \approx 8.7^2 + 5^2 \Rightarrow r^2 \approx 75.69 + 25 \Rightarrow$
   $r^2 \approx 100.69 \Rightarrow r \approx \sqrt{100.69} \approx 10.03$ cm.

45. The sides have length 14 cm, so the perimeter is $5 \cdot 14 = 70$ cm. We can find the apothem with the Pythagorean theorem: $r^2 = a^2 + \left(\frac{s}{2}\right)^2 \Rightarrow$
$11.9^2 \approx a^2 + 7^2 \Rightarrow 141.61 \approx a^2 + 49 \Rightarrow$
$92.61 \approx a^2 \Rightarrow a \approx \sqrt{92.61}$ cm. The area of a regular polygon can be found using $A = \frac{1}{2}Pa$. For this situation, $A \approx \frac{1}{2}(70)\left(\sqrt{92.61}\right) \approx 35\sqrt{92.61} \approx 336.82 \, \text{cm}^2$.

49.
   a. In the first figure, we can use x and y to represent the legs for the first figure.

   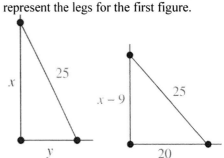

   Since the ladder is lowered 9 feet, the second figure will have $x - 9$ and 20 as the legs.
   b. For the first figure, we'll have $x^2 + y^2 = 25^2$ and from the second, we'll have
   $(x-9)^2 + 20^2 = 25^2$. Solving for x, we get
   $(x-9)^2 + 400 = 625 \Rightarrow (x-9)^2 = 225 \Rightarrow$
   $(x-9) = \sqrt{225} \Rightarrow x - 9 = 15 \Rightarrow x = 24$ ft. Now substituting that into the first equation we get
   $24^2 + y^2 = 25^2 \Rightarrow 576 + y^2 = 625 \Rightarrow$
   $y^2 = 49 \Rightarrow y = \sqrt{49} \Rightarrow y = 7$ ft. The ladder was originally 7 feet from the wall.

Section 11.4

1.
   a. There are 8 cubes here (1 is hidden), so the volume is 8 cm^3.

2.
   a. There are about 10 square units.

3.
   a. $\overline{BD}$

4.
   a. Because the radius, apothem, and half of one side form a right triangle, $r^2 = a^2 + \left(\frac{s}{2}\right)^2$.

5.
   a. $B = \frac{1}{2}Pa \Rightarrow B = \frac{1}{2}(11 \cdot 5)(7.6) = 209$ cm^2.
   b. The lateral surface area is found by finding the value of the slant height.
   $l^2 = a^2 + h^2 \Rightarrow l^2 = 7.6^2 + 15^2 \Rightarrow$
   $l^2 = 282.76 \Rightarrow l = \sqrt{282.76}$. Then the lateral surface area is $\frac{1}{2}Pl = \frac{1}{2}(55) \cdot \sqrt{282.76} \approx 462.43$ cm^2.
   c. The surface area will be:
   $S.A. = B + \frac{1}{2}Pl \approx 462.43 + 209 \approx 671.43$ cm^2.

d. The volume of the pyramid is
$V = \frac{1}{3}Bh \Rightarrow V = \frac{1}{3}(209)(15) = 1045$ cm^3.

11. If the painter needs to paint all sides except the bottom, then the surface area to be painted is
$S.A. = B + Ph = 55 \cdot 8 + (2 \cdot 55 + 2 \cdot 8) \cdot 10 = 440 + 1100 + 160 + = 1700$ ft^2. Each gallon covers 275 square feet, so the number of gallons of paint needed is $\frac{1700}{275} = 6\frac{2}{11} \approx 6.18$. But paint is sold by the gallon, so the painter needs to buy 7 gallons.

13. The surface area of a right rectangular prism is $S.A. = 2B + Ph$. The perimeter here is $P = 2l + 2w$ and the area of the base is $B = lw$. Since we know that the surface area is 856 cm^2, $h = 14$ cm, and that $l = w + 2$, we can solve for w:
$S.A. = 2lw + (2l + 2w)h \Rightarrow$
$856 = 2(w+2)(w) + (2(w+2) + 2w)(14) \Rightarrow$
$856 = 2w^2 + 4w + 28w + 56 + 28w \Rightarrow$
$0 = 2w^2 + 60w - 800 \Rightarrow 0 = w^2 + 30w - 400 \Rightarrow$
$(w-10)(w+40) = 0 \Rightarrow w - 10 = 0$ or $w + 40 = 0 \Rightarrow$
$w = 10$ or $w = -40$. Since the width can't be negative, it must be 10 cm. This means the length is 12 cm. So the volume of the prism is
$V = Bh = 12 \cdot 10 \cdot 14 = 1680$ cm^3.

17. Since 1 L = 1000 cm^3, then 52 L = 52,000 cm^3. We can use the volume formula for a prism to find the height: $V = Bh \Rightarrow$
$52,000 = (40) \cdot (35)h \Rightarrow 52,000 = 1400h \Rightarrow$
$h = \frac{52,000}{1400} = 37\frac{1}{7} \approx 37.14$ cm.

21. If the original radius r is decreased by 5%, then the new radius is $0.95r$. If the original height is h, we could call the new height H. The volume of the container would be the same for the original and the new shape:
$V = Bh = \pi \cdot r^2 h$ and $V = BH = \pi \cdot (0.95r)^2 H \Rightarrow$
$\pi \cdot r^2 h = \pi \cdot (0.95r)^2 H \Rightarrow r^2 h = (0.95)^2 r^2 H \Rightarrow$
$h = (0.95)^2 H \Rightarrow H = \frac{1}{0.95^2}h \approx 1.108h$. The new height must be about 110.8% of the original height, which is an increase of about 10.8%.

22.
a. $\left(\frac{567 \text{ in}^3}{1}\right)\left(\frac{1 \text{ ft}}{12 \text{ in}}\right)^3 = \left(\frac{567 \text{ in}^3}{1}\right)\left(\frac{1 \text{ ft}^3}{12^3 \text{ in}^3}\right) =$
$\left(\frac{567 \text{ in}^3}{1}\right)\left(\frac{1 \text{ ft}^3}{1728 \text{ in}^3}\right) = \frac{21}{64}$
$= 0.328125 \approx 0.3281$ ft^3.

b. $\left(\frac{789 \text{ in}^3}{1}\right)\left(\frac{1 \text{ ft}}{12 \text{ in}}\right)^3\left(\frac{1 \text{ yd}}{3 \text{ ft}}\right)^3 =$
$\left(\frac{789 \text{ in}^3}{1}\right)\left(\frac{1 \text{ ft}^3}{12^3 \text{ in}^3}\right)\left(\frac{1 \text{ yd}^3}{3^3 \text{ ft}^3}\right) =$
$\left(\frac{789 \text{ in}^3}{1}\right)\left(\frac{1 \text{ ft}^3}{1728 \text{ in}^3}\right)\left(\frac{1 \text{ yd}^3}{27 \text{ ft}^3}\right) \approx 0.0169$ yd^3.

23.
a. The volume displaced is the volume of a cylinder with radius 2.5 cm and height 0.25 cm, $V = Bh \Rightarrow V = \pi(3.5)^2(0.25) = 3.0625\pi \approx 9.6$ cm^3.
b. The volume of the object is equal to the volume of the water it displaces, so the volume is also approximately 9.6 cm^3.

26.
a. $\left(\frac{810.5 \text{ leeds}^3}{1}\right)\left(\frac{0.25 \text{ zippos}}{4.5 \text{ leeds}}\right)^3 =$
$\left(\frac{810.5 \text{ leeds}^3}{1}\right)\left(\frac{0.25^3 \text{ zippos}^3}{4.5^3 \text{ leeds}^3}\right) \approx$
0.13897 cubic zippos $\approx$ 0.14 cubic zippos.

27.
a. With a ratio of 1 inch of fish per gallon of water, then because a marigold swordtail fish is 5 inches long, each one would require 5 gallons of water. With 248 gallons in the aquarium, it could hold $\frac{248}{5} = 49.6$ or at most 49 marigold swordtail fish.
b. First, we can find the volume of the aquarium.
$V = Bh \Rightarrow V = (40)(15)(14) = 8400$ in^3.
Converting this to gallons:
$\left(\frac{8400 \text{ in}^3}{1}\right)\left(\frac{1 \text{ gallon}}{231 \text{ in}^3}\right) = 23\frac{4}{11}$ gallons. The
aquarium could hold $\frac{23\frac{4}{11}}{5} = 7\frac{3}{11} \approx 7.27$ or at most 7 marigold swordtail fish.

33.
a. This appears to be milliliters, which should be written mL.
b. This appears to be grams, which should be written g.
c. This appears to represent cubic inches, which should be in^3.
d. Since oz are a measure of weight, not volume, this should be written as fl oz.

35.
a. (ii) 15 – 30 gal b. (iv) 180 gal c. (v) $\frac{1}{2}$ gal

d. (vi) 1 gal   e. (i) 4 – 7 gal   f. (iii) 62,600 gal

36.
   b. The lateral surface area requires the slant height: $l^2 = a^2 + h^2$. Because the base is a square, the apothem is exactly one half of the side length.
   $l^2 = (1.5)^2 + 4^2 \Rightarrow$
   $l^2 = 2.25 + 16 \Rightarrow l^2 = 18.25 \Rightarrow l = \sqrt{18.25}$. Then the lateral surface area is
   $\frac{1}{2}Pl = \frac{1}{2}(12)(\sqrt{18.25}) \approx 25.6$ ft².

41.
   a. The area of the base requires the apothem:
   $r^2 = a^2 + \left(\frac{s}{2}\right)^2 \Rightarrow 8^2 = a^2 + 4^2 \Rightarrow a^2 = 48 \Rightarrow$
   $a = \sqrt{48}$ cm. Then $B = \frac{1}{2}Pa \Rightarrow$
   $B = \frac{1}{2}(48)(\sqrt{48}) \approx 166.28$ cm².
   b. The lateral surface area of the pyramid requires the slant height: $l^2 = a^2 + h^2 \Rightarrow$
   $l^2 = (\sqrt{48})^2 + 14^2 \Rightarrow l^2 = 244 \Rightarrow l = \sqrt{244}$ cm. Then the lateral surface area is
   $\frac{1}{2}Pl = \frac{1}{2}(48)\sqrt{244} \approx 374.9$ cm².
   c. The surface area is $S.A. = B + \frac{1}{2}Pl$
   $= 24\sqrt{48} + 24\sqrt{244} \approx 541.2$ cm².
   d. The volume of the pyramid is $V = \frac{1}{3}Bh \Rightarrow$
   $V = \frac{1}{3}(24\sqrt{48}) \cdot 14 \approx 775.96$ cm³.

44.
   a. The surface area is $S.A. = 6s^2 = 6 \cdot 3^2 = 54$ square units.
   b. After doubling the length of all sides, the surface area is $S.A. = 6s^2 = 6 \cdot 6^2 = 216$ square units.

45.
   a. The volume of the cube is $V = 4^3 = 64$ m³.
   b. After doubling the side lengths on the cube, the volume of the cube is $V = 8^3 = 512$ m³.

51. $l^2 = r^2 + h^2 \Rightarrow l^2 = 10^2 + 18^2 \Rightarrow$
$l^2 = 100 + 324 \Rightarrow l^2 = 424 \Rightarrow l = \sqrt{424} \approx 20.6$ cm.

Chapter 11 Review

2. The correct order is (ii), (iv), (i), and (iii).

4.
   b. The precision is $\frac{1}{4}$ unit, and the GPE is $\frac{1}{8}$ unit.

5.
   a. Assuming that the precision is 0.1 yd, then the GPE is 0.05 yd.
   b. Assuming that the precision is 1 ft, then the GPE is 0.5 ft.

7.
   a. Assuming precision of 0.01 pounds, then the true weight is at least 12.315 pounds and at most 12.325 pounds.

8. c. $\left(\frac{8 \text{ inks}}{1}\right)\left(\frac{3 \text{ caps}}{5 \text{ inks}}\right)\left(\frac{7 \text{ pens}}{4 \text{ caps}}\right) = \frac{42}{5} = 8\frac{2}{5}$ pens

9. a. $\left(\frac{5 \text{ in}}{1}\right)\left(\frac{1 \text{ ft}}{12 \text{ in}}\right)\left(\frac{1 \text{ mi}}{5280 \text{ ft}}\right) \approx 7.89 \cdot 10^{-5}$ mi

11. 6 caps would be equivalent to 15 inks, but also equivalent to 8 pens. Since it takes more inks of any other unit, ink is the smallest.

16.
   a. The area of a triangle can be found with $A = \frac{1}{2}bh = \frac{1}{2} \cdot 3 \cdot 4 = 6$ square units.
   b. The area of a triangle can be found with $A = \frac{1}{2}bh = \frac{1}{2} \cdot 4 \cdot 3 = 6$ square units.

19.
   a. The diameter is about 26 mm, so $C = \pi \cdot d = \pi(26) \approx 81.7$ mm².
   b. The diameter is about 30 mm, so the radius is about 15 mm. Then $A = \pi \cdot r^2 \Rightarrow$
   $A = \pi \cdot (15)^2 = 225\pi \approx 706.9$ mm².

20.
   a. If the original circumference is $C = 2\pi \cdot r$, then multiplying the radius by 4 would produce a new circumference:
   $C' = 2\pi \cdot (4r) = 4(2\pi \cdot r) = 4 \cdot C$. The new circumference is 4 times the original circumference.
   b. If the original area is $A = \pi \cdot r^2$, then multiplying the radius by 4 would produce a new area: $A' = \pi \cdot (4r)^2 = 16(\pi \cdot r^2) = 16 \cdot A$. The new area is 16 times the original area.

23.
   a. $\left(\frac{75 \text{ yd}^2}{1}\right)\left(\frac{3 \text{ ft}}{1 \text{ yd}}\right)^2 = \left(\frac{75 \text{ yd}^2}{1}\right)\left(\frac{3^2 \text{ ft}^2}{1 \text{ yd}^2}\right) =$
   $\left(\frac{75 \text{ yd}^2}{1}\right)\left(\frac{9 \text{ ft}^2}{1 \text{ yd}^2}\right) = 675$ ft².

24. All answers are rounded to two decimal places.

   b. $\left(\dfrac{4 \text{ ft}^2}{1}\right)\left(\dfrac{12 \text{ in}}{1 \text{ ft}}\right)^2\left(\dfrac{2.54 \text{ cm}}{1 \text{ in}}\right)^2 =$
   $\left(\dfrac{4 \text{ ft}^2}{1}\right)\left(\dfrac{12^2 \text{ in}^2}{1 \text{ ft}^2}\right)\left(\dfrac{2.54^2 \text{ cm}^2}{1 \text{ in}^2}\right) \approx 3716.12 \text{ cm}^2.$

29. The perimeter of the triangle is $P = 17 + 20 + 25 = 62$ cm. The perimeter of a square is $P = 4s \Rightarrow 62 = 4s \Rightarrow s = 15.5$ cm.

32.
   b. The rectangular sample has area of $A = 4 \cdot 3 = 12 \text{ cm}^2$ and has 50 mg of lead. This is a ratio of $\frac{50}{12} \approx 4.17$ mg of lead per cm². This is greater than 1, so this indicates lead based paint.

35. There are 8 boxes with a total area of 576 cm², so each box has area of $\frac{576}{8} = 72$ cm². Since each box is a square, then we know
   $A = s^2 \Rightarrow 72 = s^2 \Rightarrow s = \sqrt{72}$ cm. To find AB, we can use the Pythagorean theorem using a triangle with sides of $3s$ and $2s$:
   $(3s)^2 + (2s)^2 = AB^2 \Rightarrow AB^2 = 9s^2 + 4s^2 \Rightarrow$
   $AB^2 = 13\left(\sqrt{72}\right)^2 \Rightarrow AB^2 = 13 \cdot 72 \Rightarrow$
   $AB^2 = 936 \Rightarrow AB = \sqrt{936} \approx 30.6$ cm.

37. We could take the picture and include one more height segment from B. Then the shape is broken into 2 triangles and a rectangle.

   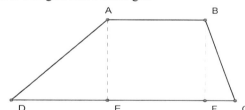

   We can use the Pythagorean theorem to find DE and CF. First, we can find DE:
   $(3.54)^2 + DE^2 = (6.25)^2 \Rightarrow DE = \sqrt{26.5309}$ cm.
   Then we can find CF:
   $(3.54)^2 + CF^2 = (4.00)^2 \Rightarrow CF = \sqrt{3.4684}$ cm.
   Then the area of the trapezoid is:
   $A = \tfrac{1}{2}(b_1 + b_2)h \Rightarrow$
   $A = \tfrac{1}{2}\left(\sqrt{26.5309} + 5 + \sqrt{3.4684} + 5\right)(3.54) \Rightarrow$
   $A \approx 30.1133 \approx 30 \text{ cm}^2.$

38.
   a. $15^2 + 18^2 = 549$ and $24^2 = 576$; since $15^2 + 18^2 < 24^2$, this is an obtuse triangle.

39.
   a. No, this cannot form a triangle because $35 \not< 2 + 15$, so these lengths fail the triangle inequality.

41. For each, find the missing angle or angles and notice that the shortest side will be opposite the smallest angle.
   a. First, find the missing angle:
   $m\angle A + m\angle B + m\angle C = 180° \Rightarrow$
   $25° + 75° + m\angle C = 180° \Rightarrow m\angle C = 80°$. Since $\angle A$ is the smallest angle, it will be opposite the shortest side $\overline{BC}$.

43.
   a. For this octagon, the sides will be found by dividing the perimeter by 8: $\frac{176}{8} = 22$ cm. The apothem of the octagon can be found using:
   $r^2 = a^2 + \left(\tfrac{s}{2}\right)^2 \Rightarrow 28.7^2 = a^2 + \left(\tfrac{22}{2}\right)^2 \Rightarrow$
   $a^2 = 702.69 \Rightarrow a = \sqrt{702.69} \approx 26.5$ cm.

47.
   a. The lateral surface area of a prism is $Ph = (5 \cdot 8.5)(20.7) = 879.75 \text{ cm}^2.$
   b. $B = \tfrac{1}{2}Pa \Rightarrow B = \tfrac{1}{2}(5 \cdot 8.5)(5.8) = 123.25 \text{ cm}^2.$

50.
   c. The slant height is found using
   $l^2 = a^2 + h^2 \Rightarrow l^2 = \left(\sqrt{9.72}\right)^2 + 8^2 \Rightarrow$
   $l^2 = 73.72 \Rightarrow l = \sqrt{73.72} \approx 8.59$ cm.
   d. The lateral surface area of a pyramid is
   $\tfrac{1}{2}Pl = \tfrac{1}{2}(21.6)\left(\sqrt{73.72}\right) = (10.8)\left(\sqrt{73.72}\right)$
   $\approx 92.73 \text{ cm}^2.$

54. The volume of the object is equal to the volume of water displaced. The volume of water displaced is the volume of a prism with height 12 cm, length 30 cm, and width 25 cm.
   $V = Bh \Rightarrow V = (30 \cdot 25) \cdot 12 = 9000 \text{ cm}^3.$

58.
   b. Since we know that the surface area is $S.A. = 6s^2$, we can solve for s:
   $1350 = 6s^2 \Rightarrow s^2 = 225 \Rightarrow s = \sqrt{225} = 15$ cm.

Chapter 11 Test

4.
   a. Clark's measurement (2.38 ft) is more precise because it measures to the nearest hundredth of a foot.
   b. We cannot tell which is more accurate without knowing the actual circumference.
   c. We might *expect* the more precise measurement to be more accurate, but as seen in previous sections that is not always the case.

6.
   a. $\left(\frac{800 \text{ yd}}{1}\right)\left(\frac{3 \text{ ft}}{1 \text{ yd}}\right)\left(\frac{1 \text{ mi}}{5,280 \text{ ft}}\right) = \frac{5}{11} \approx 0.45$ mi.

8. Measuring with a metric ruler, the radius is 20 mm.
   a. The circumference is $C = 2\pi r \Rightarrow$
   $C = 2\pi \cdot 20 = 40\pi \approx 125.6637 \approx 126$ mm.

10.
   a. Since the sector has the angle measure $72°$, then the sector will have $\frac{72}{360}$ of the area of the circle: $A = \left(\frac{72}{360}\right)\pi \cdot r^2 \Rightarrow$
   $A = \left(\frac{72}{360}\right)\pi \cdot (8)^2 = 12.8\pi \approx 40.21$ cm^2.

15. The length of the top side is 4 units, and the bottom is 8 units. The left side requires the Pythagorean theorem: $1^2 + 3^2 = c^2 \Rightarrow 10 = c^2 \Rightarrow c = \sqrt{10}$ units. The right side is $5^2 + 3^2 = d^2 \Rightarrow 34 = d^2 \Rightarrow d = \sqrt{34}$ units. Then the perimeter is $\sqrt{34} + 4 + \sqrt{10} + 8 \approx 20.993 \approx 21.0$ units.

16. Since the area of the polygon is 330 square units and there are 15 boxes, then each box must contain $\frac{330}{15} = 22$ square units. The area of each box is $A = s^2 \Rightarrow 22 = s^2 \Rightarrow s = \sqrt{22}$. Based on the picture, the legs of the right triangle with hypotenuse of $\overline{AB}$ are 6s and 2s, so the Pythagorean theorem would show:
$\left(6\sqrt{22}\right)^2 + \left(2\sqrt{22}\right)^2 = AB^2 \Rightarrow 880 = AB^2 \Rightarrow$
$AB = \sqrt{880} = 4\sqrt{55} \approx 29.66$ units.

18. We can determine the type of triangle by comparing $a^2 + b^2$ and c^2 where c is the longest side.
   a. $14^2 + 18^2 = 520$ and $28^2 = 784$. Since $14^2 + 18^2 < 28^2$, the triangle is obtuse.

19.
   b. The apothem of the polygon is $A = \frac{1}{2}Pa \Rightarrow$
   $946.37 \approx \frac{1}{2}(112)(a) \Rightarrow a \approx 16.90$ cm. We can use the Pythagorean theorem to find the radius, but we need the length of the side. Since there are 8 sides, each one is $\frac{112}{8} = 14$:
   $\left(\frac{s}{2}\right)^2 + a^2 = r^2 \Rightarrow \left(\frac{14}{2}\right)^2 + 16.90^2 \approx r^2 \Rightarrow$
   $r^2 \approx 334.61 \Rightarrow r \approx \sqrt{334.61} \approx 18.29$ cm.

22.
   a. $\left(\frac{s}{2}\right)^2 + a^2 = r^2 \Rightarrow \left(\frac{3.22}{2}\right)^2 + a^2 = 4.21^2 \Rightarrow$
   $a^2 = 15.132 \Rightarrow a = \sqrt{15.132} \approx 3.89$ cm.
   c. We can use the Pythagorean theorem to find the slant height: $h^2 + a^2 = l^2 \Rightarrow$
   $(15.03)^2 + \left(\sqrt{15.132}\right)^2 = l^2 \Rightarrow$
   $l^2 = 241.0329 \Rightarrow l = \sqrt{241.0329} \approx 15.53$ cm.

24. 28 L = 28,000 cm^3, so we can use the volume of a prism formula to find the height, h.
$V = Bh \Rightarrow 28,000 = (55 \cdot 42)h \Rightarrow h = 12\frac{4}{33} \approx 12$ cm.

Chapter 12 – Section 12.1

1.
   a. The angles are congruent because they are corresponding parts of congruent triangles (CPCTC).

2.
   a. *AEFG*

4.
   a. $\overline{QW}$, because this segment is one side of each of the angles, and therefore the included side.
   b. Triangle names can vary. This diagram implies $\triangle QWT \cong \triangle NSU$.

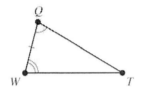
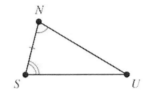

8. Here CPCFC will represent corresponding parts of congruent figures are congruent.

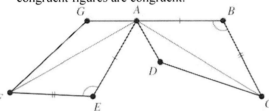

   a.

Statement	Reason
1. $ABCD \cong AEFG$	1. Given
2. $\overline{FE} \cong \overline{CB}$	2. CPCFC
3. $\angle E \cong \angle B$	3. CPCFC
4. $\overline{AE} \cong \overline{AB}$	4. CPCFC
5. $\triangle AEF \cong \triangle ABC$	5. SAS congruence axiom
6. $\overline{AF} \cong \overline{AC}$	6. CPCTC

   b.

Statement	Reason
1. $ABCD \cong AEFG$	1. Given
2. $\angle GAE \cong \angle DAB$	2. CPCFC
3. $m\angle DAG = m\angle DAE + m\angle GAE$	3. Angle addition postulate
4. $m\angle DAG = m\angle DAE + m\angle GAE$	4. Because $m\angle GAE = m\angle DAB$
5. $m\angle DAG = m\angle BAE$	5. Angle addition postulate
6. $\angle DAG \cong \angle BAE$	6. Definition of congruence

10. All sides are congruent. We need to prove *ABCD* is a parallelogram. Draw the diagonal from *A* to *C*, which creates $\overline{AC} \cong \overline{AC}$ by the reflexive property of congruence.

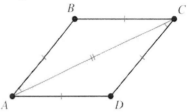

Then $\triangle ABC \cong \triangle CDA$ by the SSS congruence axiom, and $\angle BAC \cong \angle DCA$ by CPCTC. The transversal $\overrightarrow{AC}$ that cuts the lines $\overrightarrow{AB}$ and $\overrightarrow{CD}$ creates a pair of corresponding angles that are congruent, so $\overrightarrow{AB}$ and $\overrightarrow{CD}$ are parallel lines by the F Postulate. So $\overline{AB}$ and $\overline{CD}$ are parallel. Next, draw the diagonal from *B* to *D*, which creates $\overline{BD} \cong \overline{BD}$ by the reflexive property of congruence.

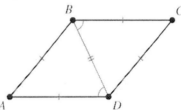

Then $\triangle ADB \cong \triangle CBD$ by the SSS congruence axiom, and $\angle ADB \cong \angle CBD$ by CPCTC. The transversal $\overrightarrow{BD}$ that cuts the lines $\overrightarrow{AD}$ and $\overrightarrow{CB}$ creates a pair of corresponding angles that are congruent, so $\overrightarrow{AD}$ and $\overrightarrow{CB}$ are parallel lines by the F Postulate. So $\overline{AD}$ and $\overline{CB}$ are parallel. Opposite sides of *ABCD* are parallel and congruent, so *ABCD* is a rhombus.

13.
   a.

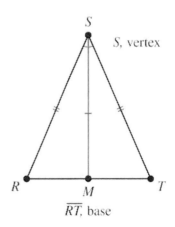

b. $\triangle RSM \cong \triangle TSM$ by the SAS congruence axiom. Then $\angle RMS \cong \angle TMS$ by CPCTC. $\angle RMS$ and $\angle TMS$ are supplementary and congruent angles, so m$\angle RMS = 90°$. Then $\overline{SM} \perp \overline{RT}$.

c. $\triangle RSM \cong \triangle TSM$ by the SAS congruence axiom. $\overline{RM} \cong \overline{TM}$ by CPCTC, so $RM = TM$. This means $RM = TM$ and M is between R and T making M the midpoint of $\overline{RT}$.

16.
  a.
  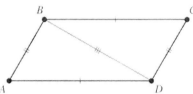
  We are given that $\overline{AB} \cong \overline{CD}$ and $\overline{AD} \cong \overline{CB}$. Also, $\overline{BD} \cong \overline{DB}$ by the reflexive property of congruence. Then $\triangle ABD \cong \triangle CBD$ by the SSS congruence axiom. Then $\angle ABD \cong \angle CDB$ by CPCTC. The transversal $\overline{BD}$ that cuts the lines $\overleftrightarrow{AB}$ and $\overleftrightarrow{CD}$ creates a pair of corresponding angles that are congruent, so $\overleftrightarrow{AB}$ and $\overleftrightarrow{CD}$ are parallel lines by the F Postulate. So $\overline{AB}$ and $\overline{CD}$ are parallel.

  b.
  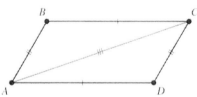
  We are given that $\overline{AB} \cong \overline{CD}$ and $\overline{BC} \cong \overline{DA}$. Also, $\overline{AC} \cong \overline{CA}$ by the reflexive property of congruence. Then $\triangle ABC \cong \triangle CDA$ by the SSS congruence axiom. Then $\angle B \cong \angle D$ by CPCTC.

17.
  a. $\overline{DA} \cong \overline{DC}$ and $\overline{BA} \cong \overline{BC}$ because all sides of a rhombus are congruent. $\overline{DB} \cong \overline{BD}$ by the reflexive property of congruence. Then $\triangle DAB \cong \triangle DCB$ by the SSS congruence axiom.

20. Suppose $\angle A \cong \angle C$. $\overline{AC} \cong \overline{CA}$ by the reflexive property of congruence. $\angle C \cong \angle A$ by the symmetric property of congruence. Then $\triangle ABC \cong \triangle CBA$ by the ASA congruence axiom. So $\overline{AB} \cong \overline{BC}$ by CPCTC. m$\angle B \neq$ m$\angle A$, so $AC \neq BC$. Then $\triangle ABC$ has exactly two congruent sides. This means $\triangle ABC$ is an isosceles triangle.

21.
  a.
  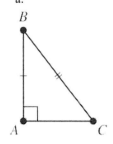

  b. $\overline{AB} \cong \overline{DE}$ and $\overline{BC} \cong \overline{EF}$ are given. $(AC)^2 + (AB)^2 = (BC)^2$ and $(DF)^2 + (DE)^2 = (EF)^2$ since they are right triangles. Then $(AC)^2 = (BC)^2 - (AB)^2$ and $(DF)^2 = (EF)^2 - (DE)^2$. But since $BC = EF$ and $AB = DE$, we can substitute:
  $(AC)^2 = (BC)^2 - (AB)^2 \Rightarrow (AC)^2 = (EF)^2 - (DE)^2$
  $\Rightarrow (AC)^2 = (DF)^2 \Rightarrow AC = DF \Rightarrow \overline{AC} \cong \overline{DF}$.
  Then $\triangle ABC \cong \triangle DEF$ by the SSS congruence axiom.

  c. The two right triangles are congruent.

27. $\overline{OH} \cong \overline{EA}$ so that you can use the ASA congruence axiom to conclude $\triangle HOT \cong \triangle AET$. Or $\overline{OT} \cong \overline{ET}$ so that we could use the AAS congruence axiom to conclude $\triangle HOT \cong \triangle AET$. Or $\overline{HT} \cong \overline{AT}$ so that we could use the AAS congruence axiom to conclude $\triangle HOT \cong \triangle AET$.

29. No, and here is a counterexample (there are many).

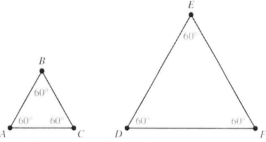

The corresponding angles of $\triangle ABC$ and $\triangle DEF$ are congruent, but $\triangle ABC \not\cong \triangle DEF$ because the corresponding sides do not have the same length.

30. No, and here is a counterexample (there are many).

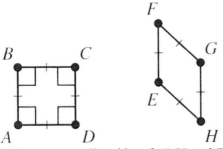

The corresponding sides of *ABCD* and *EFGH* are congruent, but $ABCD \not\cong EFGH$ because the corresponding angles are not congruent.

35.

a.

Statement	Reason
1. $\overline{AD} \cong \overline{BC}$	1. Opposite sides of a rectangle are congruent.
2. $\angle CAD \cong \angle ACB$	2. Opposite sides of a rectangle are parallel, so pairs of alternate interior angles are congruent.
3. $\angle ADB \cong \angle DBC$	3. Opposite sides of a rectangle are parallel, so pairs of alternate interior angles are congruent.
4. $\triangle AED \cong \triangle CEB$	4. ASA congruence axiom

37. a.

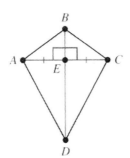

b.

Statement	Reason
1. $\overline{AE} \cong \overline{CE}$	1. $\overline{BD}$ bisects $\overline{AC}$
2. $\angle AEB \cong \angle CEB$	2. $\overline{AC} \perp \overline{BD}$
3. $\overline{BE} \cong \overline{EB}$	3. Reflexive property of congruence
4. $\triangle AEB \cong \triangle CEB$	4. SAS congruence axiom
5. $\angle ABE \cong \angle CBE$	5. CPCTC

Then $\overline{BD}$ bisects $\angle ABC$.

39. $\overline{AN} \cong \overline{AP}$ and $\overline{AT} \cong \overline{AE}$ are given. $\angle N \cong \angle P$ because $\triangle ANP$ is an isosceles triangle and the base angles are congruent. $\angle ATE \cong \angle AET$ because $\triangle ATE$ is an isosceles triangle and the base angles are congruent. Then $\angle NTA \cong \angle TEA$, because supplemental angles of congruent angles are congruent. By the AAS congruence axiom $\triangle ANT \cong \triangle APE$.

43.

a. $\overline{ON} \cong \overline{ET}$, $\overline{OW} \cong \overline{EA}$, and $\overline{DN} \cong \overline{ST}$ are given. Since W and A are midpoints, we know $DN = 2 \cdot WN$ and $ST = 2 \cdot AT$. Then $\overline{DN} \cong \overline{ST} \Rightarrow DN = ST \Rightarrow 2 \cdot WN = 2 \cdot AT \Rightarrow WN = AT \Rightarrow \overline{WN} \cong \overline{AT}$. Then $\triangle NOW \cong \triangle TEA$ by the SSS congruence axiom.

b.

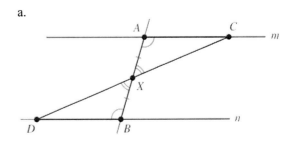

$\triangle NOW \cong \triangle TEA$ by part (a). $\angle N \cong \angle T$ by CPCTC, and $\overline{DN} \cong \overline{ST}$ and $\overline{ON} \cong \overline{ET}$ are given. Then $\triangle DON \cong \triangle SET$ by the SAS congruence axiom.

45.

a.

b.

Statement	Reason
1. $m \parallel n$	1. Given
2. $\overline{AX} \cong \overline{BX}$	2. Definition of midpoint
3. $\angle A \cong \angle B$	3. Transversal cuts parallel lines creating congruent alternate interior angles.
4. $\angle AXC \cong \angle BXD$	4. Definition of vertical angles
5. $\triangle AXC \cong \triangle BXD$	5. ASA congruence axiom

48.

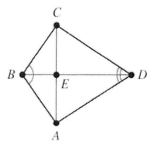

Let E be the point of intersection of the diagonals.

Statement	Reason
1. $\angle CBD \cong \angle ABD$	1. Since $\overline{BD}$ bisects $\angle B$
2. $\angle CDB \cong \angle ADB$	2. Since $\overline{BD}$ bisects $\angle D$
3. $\overline{BD} \cong \overline{BD}$	3. Reflexive property of congruence
4. $\triangle CBD \cong \triangle ABD$	4. ASA congruence axiom
5. $\overline{CB} \cong \overline{AB}$	5. CPCTC
6. $\overline{BE} \cong \overline{BE}$	6. Reflexive property of congruence
7. $\triangle CBE \cong \triangle ABE$	7. SAS congruence axiom
8. $\angle BEC \cong \angle BEA$	8. CPCTC
9. $m\angle BEC = 90°$	9. $\angle BEC$ and $\angle BEA$ are congruent and supplementary angles, so they are both right angles.

This proves that the diagonals are perpendicular.

Section 12.2

1.
  a. A and B are the given points.
  b. The constructed objects are the line segment through A and B, the arc with center B, and point C.

3.
  a. The point P.
  b. The vertical line m.

5. Answers vary.
  b.

7.
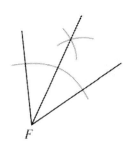

10. Extend the given line segment using a straightedge. Then set the compass opening to radius AB. Place the center point at B and swing an arc with radius AB so that it intersects the line segment. Label the intersection C. Construct the perpendicular bisector of $\overline{BD}$. Label the point of intersection of the perpendicular bisector and $\overline{BC}$ as D. Then D is the midpoint of $\overline{BC}$. Then $AD = AB + BD = 3 + 1.5 = 4.5$. Then $\overline{AD}$ has length 4.5 units.

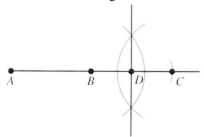

13. Construct a line segment $\overline{AB}$ using the straightedge. Construct the perpendicular bisector of $\overline{AB}$ to create a right angle. Bisect the right angle to create two 45° angles. $\angle NMB$ has measure 45° as shown.

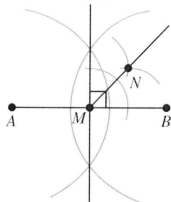

17.
  a. Create a right angle $\angle A$, as described in the answer to Question 8, by constructing the perpendicular bisector of a line segment. Then create an angle bisector of $\angle A$.

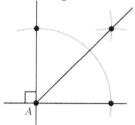

Set the compass opening to r. Put the center point at A and swing an arc with radius r that intersects the angle bisector. Label the point of intersection C.

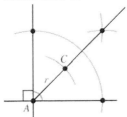

Construct a line through C that is perpendicular to a side of ∠A. Label the point of intersection B. Construct a line through C that is perpendicular to the other side of ∠A. Label the point of intersection D. Then ABCD is a square.

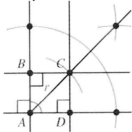

  b. Set the compass radius to AB and place the compass points on consecutive vertices. The sides of ABCD have the same length.

19. Answers vary.

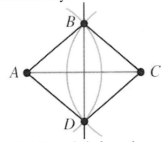

  a. A right angle is formed.
  b. The diagonals of a rhombus are perpendicular, and the diagonals bisect each other. Construct the perpendicular bisector of $\overline{AC}$. Set the radius of the compass to r units that is greater than $\frac{AC}{2}$. Place the center point of the compass at A and swing an arc with radius r. Maintaining the same opening of the compass, place the center point of the compass at C and swing an arc. Let B and D be the points of intersection of the two arcs as shown. Then ABCD is a rhombus.

21.

  a. Suppose C lies on the perpendicular bisector of $\overline{AB}$.

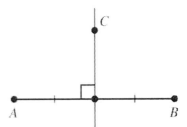

Let M be the midpoint of $\overline{AB}$, and draw in the segments to form △ABC.

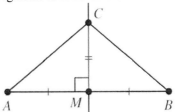

From the diagrams, $\overline{AM} \cong \overline{BM}$, ∠AMC ≅ ∠BMC, and $\overline{MC} \cong \overline{MC}$. Then △AMC ≅ △BMC by the SAS congruence axiom. Then $\overline{AC} \cong \overline{BC}$ by CPCTC.

23.
  a. The diagonals are congruent and perpendicular.

26.
  a.

  b. Use a straightedge to construct chords $\overline{AB}$ and $\overline{BC}$, then use the techniques for constructing the perpendicular bisector of a line segment. The center of the circle lies on the perpendicular bisector of a chord. The point that lies on the intersection of two or more perpendicular bisectors of chords is the center of the circle.

28. Use the straightedge to construct a line segment, then pick a point A on the line segment. Set the compass radius to 2 units. Put the center point at A and swing an arc that intersects the line segment. Label the point of intersection B. Then AB = 2 units. Maintaining the same compass opening, put the center point at B and swing another arc that intersects the line segment. Label the point of intersection C. Then BC = 2 units. Use a construction technique to construct the perpendicular bisector of $\overline{BC}$. Label the midpoint of $\overline{BC}$ as D. Then $AD = 3$ units.

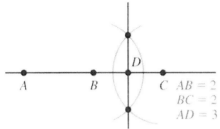

Set the compass radius to 2 units. Place the center point of the compass at D and swing an arc. Maintaining the same compass opening, place the center point of the compass at A and swing an arc. Label the intersection of the arcs E.

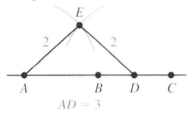

Then $\triangle EAD$ is a triangle with sides having the ratio 2:2:3.

32. Put the center point on X and swing an arc so that it intersects both sides of $\angle X$. Label the points of intersection A and B. Use a straightedge to draw the rays $\overrightarrow{XA}$ and $\overrightarrow{XB}$. Set the compass radius to a units. Put the center point of the compass at X and swing an arc so that it intersects $\overrightarrow{XA}$. Label the point of intersection C. Then $XC = a$ units. Set the compass radius to b units. Put the center point of the compass at X and swing an arc so that it intersects $\overrightarrow{XB}$. Label the point of intersection D. Then $XD = b$ units. Use a construction technique to draw a line through C and parallel to $\overrightarrow{XD}$, and a line through D that is parallel to $\overrightarrow{XC}$. Let the intersection of these constructed lines be E. Then $\overline{XC} \parallel \overline{DE}$ and $\overline{XD} \parallel \overline{CE}$, making quadrilateral $XDEC$ is a parallelogram.

33. $140 = 2^2 \cdot 5 \cdot 7$, meaning 140 has the prime 7 as a factor, and 7 is not a Fermat prime number. So Euclidean tools cannot be used to construct a regular polygon with 140 sides.

39. Use a straightedge to extend $\overline{AB}$. Construct a line that goes through A and is perpendicular to $\overline{AB}$. Put the center point of the compass at B and swing an arc so that it intersects the perpendicular line in two locations. Label the points of intersections C and D. Then $\overline{CD} \perp \overline{AB}$ and $BC = BD$. Then

$\triangle BCD$ is an isosceles triangle with apex B and altitude $\overline{AB}$.

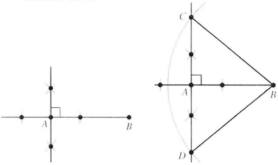

41. Construct a line that goes through A and is perpendicular to a side of $\angle A$, as this will create a 40° angle. Bisect the 40° angle to create two 20° angles.

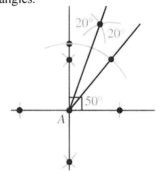

44. Use a straightedge to construct a line segment. Pick a point A on the line segment. Set the compass radius to 1 unit. Put the center point at A and swing an arc that intersects the line segment. Label the point of intersection B. Repeat this as shown in the diagram so that $AD = 3$ units. Pick a point E on the line segment such that A is between E and D, as shown. Construct a line through E that is perpendicular to $\overline{AD}$. Set the compass radius to AC, then use the compass to construct a point G on the perpendicular line such that $EG = 2$. Then $\triangle ADG$ is an obtuse triangle with a base of 3 units, a height of 2 units, and an area of $A = \frac{1}{2}bh = \frac{1}{2} \cdot 3 \cdot 2 = 3$ square units.

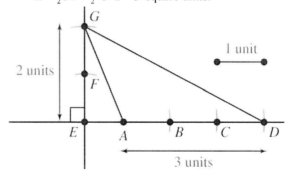

47. Construct an equilateral triangle. Bisect an interior angle to create a 30° angle. Bisect a 30° angle to obtain a 15° angle. Copy a 15° angle so that one of the sides of the angles is a side of the triangle with a common vertex. Then 60° + 15° = 75°, so ∠ABC has measure 75°.

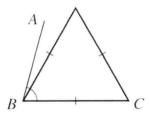

54.

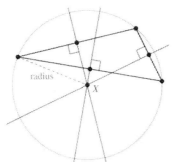

Create the perpendicular bisectors and find their intersection point. With the center point at the intersection point *X*, create the radius from *X* to any of the original vertices. Construct the circle with the radius in the compass and it will intersect all of the vertices.

55.
 a. *X* lies on the perpendicular bisector of $\overline{AB}$, so *AX* = *BX*.
 b. *X* lies on the perpendicular bisector of $\overline{BC}$, so *BX* = *CX*.
 c. *AX* = *BX* and *BX* = *CX* from part (a) and (b). This means $\overline{AX} \cong \overline{BX}$ and $\overline{BX} \cong \overline{CX}$, so $\overline{AX} \cong \overline{CX}$ by the transitive property of congruence. Then *AX* = *CX*.
 d. *AX* = *BX* = *CX*. This means that points *A*, *B*, and *C* lie on a circle with center *X* and radius *AX*. So the circumcenter *X* of △*ABC* is equidistant from vertices *A*, *B*, and *C*.

62. The diagonals of the square are also diameters of the circle. Use a straightedge to construct a diameter of the circle. Label the endpoints *A* and *C*. Construct a perpendicular bisector of the diameter. Label the points of intersection of the perpendicular line *B* and *D*. Then *ABCD* is a square that is inscribed in the circle.

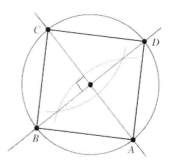

Section 12.3

1. No, the triangles are not similar because corresponding sides are not proportionate: $\frac{5}{4} \neq \frac{4}{3}$.

3.
 a.
 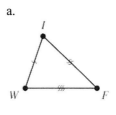 

 b. △*WIF* ~ △*OHA*
 c. SSS similarity axiom

8. Answers vary. △*ABC* ~ △*DBE*

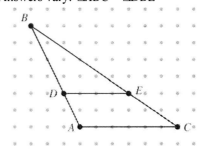

11.
 a.
 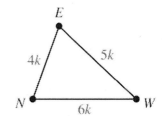

 b. Since the perimeter of △*NEW* is 48 units, then we have the equation
 $6k + 5k + 4k = 48 \Rightarrow 15k = 48 \Rightarrow k = \frac{48}{15} = \frac{16}{5}$.
 Then $NE = 4k = 4\left(\frac{16}{5}\right) = 12.8$ units,
 $EW = 5k = 5\left(\frac{16}{5}\right) = 16$ units, and
 $NW = 6k = 6\left(\frac{16}{5}\right) = 19.2$ units.

106

13. We can set up a proportion from the picture and solve for AB.
$\frac{15}{AB} = \frac{23}{AB+7} \Rightarrow 23AB = 15(AB+7) \Rightarrow$
$23AB = 15AB + 105 \Rightarrow 8AB = 105 \Rightarrow$
$AB = 13.125$ ft.

19. Suppose $\frac{AB}{BC} = \frac{DE}{EF}$. Then $\frac{AB}{AC-AB} = \frac{DE}{DF-DE} \Rightarrow$
$AB(DF - DE) = DE(AC - AB) \Rightarrow$
$AB \cdot DF - AB \cdot DE = DE \cdot AC - DE \cdot AB \Rightarrow$
$AB \cdot DF = DE \cdot AC \Rightarrow \frac{AB}{AC} = \frac{DE}{DF}$.

25.
  a. Using the Pythagorean theorem,
  $3^2 + 10^2 = AC^2 \Rightarrow 109 = AC^2 \Rightarrow$
  $AC = \sqrt{109} \approx 10.4$ units.

26. $\frac{AB}{DE} = 5 \Rightarrow AB = 5 \cdot DE$.
  a.
  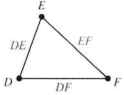 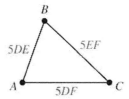

  b. $\frac{P_A}{P_D} = \frac{AB+BC+AC}{DE+EF+DF} = \frac{5 \cdot DE + 5 \cdot EF + 5 \cdot DF}{DE + EF + DF}$
  $= \frac{5 \cdot (DE + EF + DF)}{(DE + EF + DF)} = 5$. This means
  $\frac{P_A}{P_D} = 5 \Rightarrow P_A = 5 \cdot P_D$, which means the perimeter of $\triangle ABC$ is 5 times the perimeter of $\triangle DEF$.

36.
  a. $\triangle YEP \sim \triangle YXZ$
  b.
  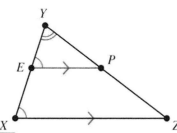

  $\overline{YX}$ is a transversal that cuts two parallel line segments $\overline{EP}$ and $\overline{XZ}$, so it creates corresponding angles that are congruent ($\angle E \cong \angle X$ as marked). $\angle Y \cong \angle Y$ by the reflexive property of congruence. So $\triangle YEP \sim \triangle YXZ$ by the AA similarity axiom.

38.
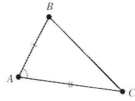 

41.
  a. $\triangle ACE \sim \triangle BCD$
  b.

Statement	Reason
1. $\angle CBD \cong \angle CAE$	1. Right angles are congruent
2. $\angle C \cong \angle C$	2. Reflexive property of congruence
3. $\triangle ACE \sim \triangle BCD$	3. AA similarity axiom

42. The ratio of two corresponding sides is $\frac{NW}{OD} = \frac{8}{6} = \frac{4}{3}$, so this is also the ratio of the heights of the triangles. If the bases of the triangles are $\overline{OD}$ and $\overline{NW}$, then we could call the heights h_O and h_N and the areas A_O and A_N. We would know that $\frac{h_N}{h_O} = \frac{4}{3} \Rightarrow h_O = \frac{3}{4} h_N$. Then we know $A_N = \frac{1}{2} \cdot 8 \cdot h_N \Rightarrow 720 = 4 \cdot h_N \Rightarrow h_N = 180$. Then $h_O = \frac{3}{4}(180) = 135$. This means $A_O = \frac{1}{2} \cdot 6 \cdot 135 = 405$ square units. We could also have used the square of the ratios to find the missing area: $\frac{A_N}{A_O} = \left(\frac{4}{3}\right)^2 \Rightarrow$
$720 = \frac{16}{9} \cdot A_O \Rightarrow A_O = \frac{720 \cdot 9}{16} = 405$ square units.

46.
  a. $\triangle HOT \sim \triangle AET$, because of the AA similarity axiom.
  b. First, we need to find one missing side so we can create a proportion. In $\triangle AET$, we know that $9^2 + ET^2 = 18^2 \Rightarrow$
  $ET = \sqrt{324 - 81} = \sqrt{243} = 9\sqrt{3}$. Then we know
  $\frac{10}{18} = \frac{n}{9\sqrt{3}} \Rightarrow 18n = 90\sqrt{3} \Rightarrow n = 5\sqrt{3} \approx 8.66$.

50.
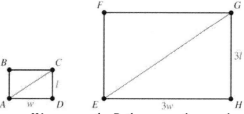

We can use the Pythagorean theorem here:
$(EG)^2 = (EH)^2 + (HG)^2 \Rightarrow$

$(EG)^2 = (3 \cdot AD)^2 + (3 \cdot DC)^2 \Rightarrow$
$(EG)^2 = 9(AD)^2 + 9(DC)^2 \Rightarrow$
$(EG)^2 = 9((AD)^2 + (DC)^2) \Rightarrow$
$(EG)^2 = 9 \cdot (AC)^2 \Rightarrow EG = 3 \cdot AC$. The diagonal $\overline{EG}$ is 3 times longer than the diagonal $\overline{AC}$.

53. a. No, the ratios of the sides are not identical: $\frac{4}{3} \neq \frac{12}{8}$ because $32 \neq 36$.
    a. Yes, because the ratios of sides are identical: $\frac{4}{3} = \frac{4}{3}$.

55. The ratio of the areas will be the square of the ratio of the sides. We can call W the area of $FGHIJ$, then $\frac{180}{W} = \left(\frac{6}{1}\right)^2 \Rightarrow 36W = 180 \Rightarrow W = 5$ cm^2.

58. Since $\triangle DOG \sim \triangle RUN$, $\frac{DO}{RU} = \frac{OG}{UN} = \frac{DG}{RN}$. In this case, this ratio is $\frac{DO}{RU} = \frac{15}{5} = 3$, so
$\frac{OG}{UN} = 3 \Rightarrow \frac{OG}{11} = 3 \Rightarrow OG = 33$ and
$\frac{DG}{RN} = 3 \Rightarrow \frac{21}{RN} = 3 \Rightarrow RN = 7$. The perimeter of $\triangle DOG$ is $15 + 21 + 33 = 69$ units and the perimeter of $\triangle RUN$ is $5 + 7 + 11 = 23$ units.

Chapter 12 Review

3. a. $\angle L$ b. $\angle KLW$

4.
   a. The two triangles are congruent so their corresponding angles are congruent (CPCTC).
   b. The two triangles are congruent, so their corresponding sides are congruent (CPCTC).

6. $\overline{OC}$ and $\overline{OD}$ equal to the radius of the circle, and $\overline{DC}$ is longer than the radius. So $\triangle DOC$ is an isosceles triangle with base angles $\angle OCE$ and $\angle ODE$. The base angles of an isosceles triangle are congruent, so $\angle OCE \cong \angle ODE$.

8. The area of a parallelogram is $A = bh \Rightarrow 360 = 30 \cdot BE \Rightarrow BE = 12$. Then using the Pythagorean theorem on $\triangle BAE$,
$12^2 + AE^2 = 20^2 \Rightarrow AE^2 = 256 \Rightarrow$
$AE = \sqrt{256} = 16$ units. Since $AE = 16$ units and $AD = 30$ units, we know that $ED = 14$ units.

12.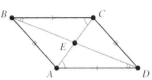

    a. $ABCD$ is a parallelogram. $\overline{BC} \cong \overline{DA}$ opposite sides of a parallelogram are congruent. Then $\angle BCA \cong \angle DAC$ and $\angle CBE \cong \angle ADE$, because a transversal that cuts parallel lines creates congruent alternate interior angles. So $\triangle BCE \cong \triangle DAE$ by the ASA congruence axiom.
    b. From part (a), $\triangle BCE \cong \triangle DAE$ so $\overline{BE} \cong \overline{ED}$ by CPCTC. The intersection of the diagonals, point E, is the midpoint of diagonal $\overline{BD}$.
    c. From part (a), $\triangle BCE \cong \triangle DAE$ so $\overline{CE} \cong \overline{AE}$ by CPCTC. So the intersection of the diagonals, point E, is the midpoint of diagonal $\overline{AC}$.

15. a.

Statement	Reason
1. $\angle BCE \cong \angle DAE$	1. Alternate interior angles created by a transversal that cuts parallel lines are congruent.
2. $\overline{BC} \cong \overline{DA}$	2. Sides of a rhombus are congruent
3. $\angle CBE \cong \angle ADE$	3. Alternate interior angles created by a transversal that cuts parallel lines are congruent.
4. $\triangle BCE \cong \triangle DAE$	4. ASA congruence axiom

17. Use the straightedge to construct a line. Pick a point on the line, label the point A. Set the compass radius to a. Place the center point on A and construct a circle. Label one of the points of intersection of the circle and line D. Pick a point on the circle and label it B. Maintaining the same compass opening, put the center point at D and construct an arc; then put the center point at B and swing an arc that intersects the previous arc. Label the point of intersection C. $ABCD$ is a parallelogram because opposite sides are congruent (see Example 12.9). All sides have the same length, which make the parallelogram a rhombus.

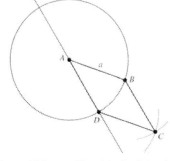

21. a. Use a straightedge to construct a line. Pick a point on the line and label it *A*. Set the radius of the compass to *b*. Place the center point at *A* and swing an arc that intersects the line at one place. Label the points of intersection *B*. Set the radius of the compass to *a*. Place the center point at *A* and swing an arc. Set the radius of the compass to *c*. Place the center point at *B* and swing an arc that intersects the previous arc. Label the point of intersection *C*. Then $\triangle ABC$ is a triangle with sides that have the lengths *a*, *b*, and *c* units.

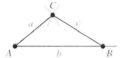

23. The center of the circle is the intersection of the perpendicular bisectors of the sides. Construct the perpendicular bisectors for two sides of the polygon. Label the intersection of a side and perpendicular bisector *A*. Label the point of intersection of the perpendicular bisectors *B*. Use a straightedge to construct a line segment from *A* to *B*. Then $\overline{AB}$ is an apothem of the hexagon.

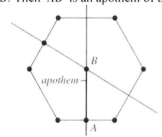

27. The known Fermat prime numbers are 3, 5, 17, 257, and 65,537.
    a. Yes, because the primes 3 and 5 appear only once in the prime factorization (2 can appear many times).
    b. No, because the prime factor of 3 appears too many times (can only appear once meaning it has an exponent of 1 at most).

31.
    a.
    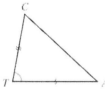
    b. $\triangle CAT \sim \triangle SUB$
    c. Two pairs of corresponding sides have the same ratio and their included angles are congruent, so the two triangles are similar by the SAS similarity axiom.

36. We can use the Pythagorean theorem:
$(AC)^2 = (AD)^2 + (CD)^2 \Rightarrow$
$(AC)^2 = (5 \cdot WZ)^2 + (5 \cdot YZ)^2 \Rightarrow$
$(AC)^2 = 25 \cdot (WZ)^2 + 25 \cdot (YZ)^2 \Rightarrow$
$(AC)^2 = 25 \cdot ((WZ)^2 + (YZ)^2) \Rightarrow$
$(AC)^2 = 25 \cdot (WY)^2 \Rightarrow AC = 5 \cdot WY$.

The diagonal $\overline{AC}$ is 5 times longer than the diagonal $\overline{WY}$.

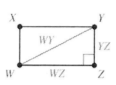 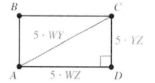

41. k, k^2

44. The ratio of the areas will be the square of the ratio of the sides. Since we know the area of $\triangle XYZ$ is $\frac{1}{2}(3.4)(14.7) = 24.99 \text{ ft}^2$, we can let the area of $\triangle ABC$ be *W*, then $\frac{W}{24.99} = \left(\frac{8.4}{14.7}\right)^2 \Rightarrow$
$W = 24.99 \cdot \left(\frac{8.4}{14.7}\right)^2 = 8.16 \text{ ft}^2$.

47. a.
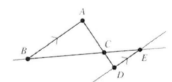

Create a line parallel to $\overline{AB}$ that passes through *D*. Then extend $\overline{BC}$ until it hits your new line, forming point *E*.

Chapter 12 Test

2. $\triangle PQR$ is an isosceles triangle, so its legs are congruent and its base angles are congruent, $\angle P \cong \angle R$. So we have $\overline{PQ} \cong \overline{RQ}$ and $\angle PQS \cong \angle RQT$. Then $\triangle PQS \cong \triangle RQT$ by the ASA congruence axiom. Then $\overline{QS} \cong \overline{QT}$ by CPCTC.

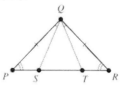

4. $\triangle LET$ is an isosceles triangle because $\overline{LE}$ and $\overline{ET}$ are both a radius of the circle. Since the base angles

are congruent, we know
$\angle ELT \cong \angle ETL \Rightarrow m\angle 4 = m\angle 1$. Then we can find these measures ΔLET using the sum of the interior angles: $m\angle 4 + m\angle 1 + 123° = 180° \Rightarrow$
$2(m\angle 4) = 57° \Rightarrow m\angle 4 = 28.5° = m\angle 1$. Because of the two supplementary angles meeting at point E, we know $m\angle 5 = 180° - 123° = 57°$. ΔMEL is an isosceles triangle because $\overline{ME}$ and $\overline{LE}$ are both a radius of the circle. Since the base angles are congruent, we know $\angle 2 \cong \angle 3$, and we can find their measure with the sum of the interior angles in ΔTHA: $m\angle 2 + m\angle 3 + 57° = 180° \Rightarrow$
$2(m\angle 3) = 123° \Rightarrow m\angle 3 = 61.5° = m\angle 2$.

6.

Statement	Reason
1. $\overline{KI} \cong \overline{IT}$	1. Given
2. $\overline{KE} \cong \overline{TE}$	2. Given
3. $\overline{IE} \cong \overline{IE}$	3. Reflexive property of congruence
4. $\Delta IKE \cong \Delta ITE$	4. SSS congruence axiom
5. $\angle K \cong \angle T$	5. CPCTC

10. Construct a line through the vertex of the angle that is perpendicular to one of the sides of the angle. This will create complementary angles, with measures 24° and 66°. Bisect the 66° angle to obtain two angles with measure 33°.

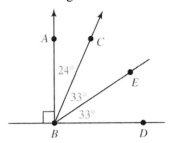

14.
 a. $\Delta ACB \sim \Delta RPM$, so $\angle A \cong \angle R \Rightarrow m\angle A = 45°$.
 b. We can set up a proportion to find the missing piece:
 $\frac{AC}{4.96} = \frac{3.60}{5.76} \Rightarrow (5.76)(AC) = (3.60)(4.96) \Rightarrow$
 $AC = \frac{(3.60)(4.96)}{5.76} = 3.1$ units.

18. Let's call P the perimeter of ΔHEN. Since the ratio of perimeters of similar objects is the same as the ratio of their corresponding sides, then we know $\frac{72}{P} = \frac{12}{1} \Rightarrow 12 \cdot P = 72 \Rightarrow P = 6$ cm.

21.
 a.

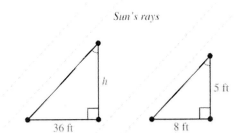

 b. The AA similarity axiom allows this conclusion because the right angles are congruent and the top angles (formed by the sun's rays) are also congruent.
 c. We can set up a proportion to find h:
 $\frac{h}{5} = \frac{36}{8} \Rightarrow 8 \cdot h = 180 \Rightarrow h = 22.5$ ft.

Chapter 13 – Section 13.1

2. The *x*-intercept is (3, 0) and the *y*-intercept is (0, –6). The slope is $\frac{-6-0}{0-3} = \frac{-6}{-3} = 2$.

4.
   a. The slope is –24, and means 24 fewer tickets are sold each day after the release of the movie.

5.
   a. There are two points (*d*, *p*) given to us in the problem: (0, 15) and (11, 20). The slope between these is $\frac{20-15}{11-0} = \frac{5}{11}$. So the equation would be $p(d) = \frac{5}{11}d + 15$.

7. Since the two points are solutions to the equation, we can create two equations: $y_1 = mx_1 + b$ and $y_2 = mx_2 + b$. Then subtracting the two equations will give us:
   $y_2 - y_1 = (mx_2 + b) - (mx_1 + b) \Rightarrow$
   $y_2 - y_1 = mx_2 + b - mx_1 - b \Rightarrow$
   $y_2 - y_1 = mx_2 - mx_1 \Rightarrow$
   $y_2 - y_1 = m(x_2 - x_1) \Rightarrow \frac{y_2 - y_1}{x_2 - x_1} = m$.

9. To use the definition of slope, we need two points. Choose any two points on the line and then use the definition.
   a. Let *x* = 1, then $y = 4 \cdot 1 + 5 = 9$. Let *x* = 2, then $y = 4 \cdot 2 + 5 = 13$. Using the definition of slope, $m = \frac{13-9}{2-1} = \frac{4}{1} = 4$.

11.
   a. The slopes are *m* = 5 and *m* = 10; these lines are **neither** parallel nor perpendicular.

13. A line parallel to $y = 3x + 4$ will have slope of $m = 3$. Our line has the form $y = 3x + b$. Since it must also pass through (5, 2), we have $2 = 3 \cdot 5 + b \Rightarrow b = 2 - 15 = -13$. The equation would be $y = 3x - 13$.

16.
   a. The position of a term.
   b. The *n*th term.

17. The slope is 7, so $\frac{\text{change in } y}{\text{change in } x} = 7$. Then $\frac{\text{change in } y}{2} = 7$. Then change in *y* = 2 · 7 = 14. The *y* value will increase by 14 units.

21. The slope of this line would be $m = \frac{k-w}{4-4} = \frac{k-w}{0}$, which is undefined. So this line would be vertical and the equation could be written as $x = 4$.

25.
   a.
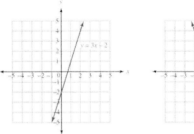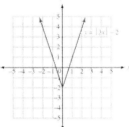

   b.
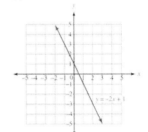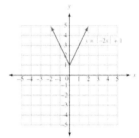

   c. The absolute value turns the line into a graph with a V-shape that bends at the *y*-intercept.

29.
   a. $y = -3(3) + 5 \Rightarrow y = -9 + 5 = -4$, so the point (3, 4) is not a solution.

31.
   a. We need to check both equations: $3y + 2x = 3(8) + 2(-3) = 24 - 6 = 18$, so the point is a solution to the first equation. $2x^2 + 5y = 2(-3)^2 + 5(8) = 18 + 40 = 58$. The point is not a solution to the second equation, so it is not a solution to both equations.

35.
   a. A line through the points would have slope of $m = \frac{-9-3}{-1-5} = \frac{-12}{-6} = 2$. Our line will look like $y = 2x + b$. Since it must also pass through (5, 3), we have $3 = 2 \cdot 5 + b \Rightarrow b = 3 - 10 = -7$. The equation would be $y = 2x - 7$.

37.
   a. To find the *y*-intercept, let *x* = 0. $y = -4 \cdot 0 + 7 = 7$, so the *y*-intercept is (0, 7). To find the *x*-intercept, let *y* = 0.

0 = -4·x + 7 ⇒ -4x = -7 ⇒ x = 7/4, so the x-intercept is (7/4, 0).

b. To find the y-intercept, let x = 0. 3·0 - 4y = 24 ⇒ -4y = 24 ⇒ y = -6, so the y-intercept is (0, -6). To find the x-intercept, let y = 0. 3x - 4·0 = 24 ⇒ 3x = 24 ⇒ x = 8, so the x-intercept is (8, 0).

c. To find the y-intercept, let x = 0. y = 4, so the y-intercept is (0, 4). To find the x-intercept, let y = 0. But the equation of the line is y = 4, so this can't happen meaning there is no x-intercept.

41.
a. The slope is $5 = \frac{u-3}{4-2} \Rightarrow 10 = u - 3 \Rightarrow u = 13$.

42.
a. For each distance in miles *m* (rounded up to the nearest fifth of a mile), there will be 5*m* fifths of a mile and each one will cost $0.34. There is also the initial fee of $2.00 for a total fare of $f(m) = 0.34(5m) + 2 \Rightarrow f(m) = 1.7m + 2$ dollars.

46.
a. When *x* = *k*, the oblique line becomes y = mk + b. So the point on both lines would be (k, mk + b).

51.

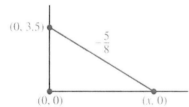

The diagram shows a representation of the situation. Solve the equation $-\frac{5}{8} = \frac{3.5-0}{0-x} \Rightarrow 5x = 8(3.5) \Rightarrow x = \frac{28}{5} = 5.6$ ft.

Section 13.2

1. Point D is not given, but we can find it by looking at the location of B and C. Point C is exactly 8 units right of point B, so point D will be exactly 8 units right of point A. The final coordinates are A(-3,-1), B(-5,4), C(3,4), and D(5,-1).

5. First, we can find the radius of the circle AB.
$AB = \sqrt{(4-(-1))^2 + (3-5)^2} \Rightarrow$
$AB = \sqrt{(4+1)^2 + (3-5)^2} \Rightarrow$
$AB = \sqrt{5^2 + (-2)^2} \Rightarrow AB = \sqrt{25+4} = \sqrt{29}$. With the center of (4, 3) and a radius of $\sqrt{29}$, the equation of the circle is $(x-4)^2 + (y-3)^2 = (\sqrt{29})^2$.

7. To get to point B from point C, we need to go left by exactly BC units. The coordinates of B are $(h - BC, k)$. The horizontal distance from A to B is $h - BC$, and since the trapezoid is isosceles, then the horizontal distance from C to D is also $h - BC$. Adding this horizontal distance to the x-coordinate of C will get us to the x-coordinate of D. So the coordinates of D are $(h + h - BC, 0) = (2h - BC, 0)$.

9.
a.

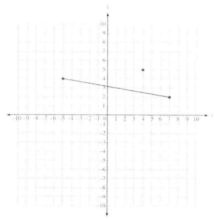

b. First, we can find the slope of the line between L and N: $m = \frac{4-2}{-5-7} = \frac{2}{-12} = -\frac{1}{6}$. The slope of the line we want will be 6, so our line will look like $y = 6x + b$. Since it must also pass through (4, 5), we have $5 = 6·4 + b \Rightarrow b = 5 - 24 = -19$. The equation would be $y = 6x - 19$.

11. We could graph the shape to see what it is, and from the graph we can see that it is a parallelogram because the opposite sides have the same slopes. The base of this parallelogram is 7 units and the height is 3 units, so the area is A = bh = 7·3 = 21 units.

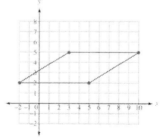

13. We can find the midpoints of the two diagonals. The midpoint of $\overline{AC}$ is $\left(\frac{-2+3}{2},\frac{4-2}{2}\right)=\left(\frac{1}{2},1\right)$, and the midpoint of $\overline{BD}$ is $\left(\frac{-2+3}{2},\frac{7-5}{2}\right)=\left(\frac{1}{2},1\right)$. Since the midpoints are the same, the diagonals bisect each other.

15. $\overline{AB}$ and $\overline{CD}$ are opposite sides; the slope of $\overline{AB}$ is $\frac{4}{6}$, and the slope of $\overline{CD}$ is $\frac{4}{6}$. $\overline{BC}$ and $\overline{AD}$ are opposite sides; the slope of $\overline{BC}$ is 2 and the slope of $\overline{AD}$ is 2. The opposites sides of the quadrilateral are parallel, so $ABCD$ is a parallelogram.

20.
   a. $AC=\sqrt{(8-3)^2+(-1-(-2))^2} \Rightarrow$
   $AC=\sqrt{(5)^2+(1)^2}=\sqrt{25+1}=\sqrt{26}$.
   $TA=\sqrt{(10-8)^2+(7-(-1))^2} \Rightarrow$
   $TA=\sqrt{(2)^2+(8)^2}=\sqrt{4+64}=\sqrt{68}$.
   $CT=\sqrt{(10-3)^2+(7-(-2))^2} \Rightarrow$
   $CT=\sqrt{(7)^2+(9)^2}=\sqrt{49+81}=\sqrt{130}$.
   Based on these lengths, we can see $\overline{CT}$ is the longest side, so $\angle A$ is the largest angle.
   b. Based on the lengths in part (a), $\overline{AC}$ is the shortest side, so $\angle T$ has the smallest angle.

24. The slope of $\overline{AB}$ is $\frac{-4}{3}$ and the slope of $\overline{BC}$ is $\frac{3}{4}$. So product of the slopes $\overline{AB}$ and $\overline{BC}$ is -1, thus $\overline{AB}\perp\overline{BC}$. So $\triangle ABC$ is a right triangle.

27. $UF=\sqrt{(-8-7)^2+(2-5)^2} \Rightarrow$
   $UF=\sqrt{(-15)^2+(-3)^2}=\sqrt{225+9}=\sqrt{234}$.
   $FN=\sqrt{(7-3)^2+(1-5)^2} \Rightarrow$
   $FN=\sqrt{(4)^2+(-4)^2}=\sqrt{16+16}=\sqrt{32}$.
   $UN=\sqrt{(-8-3)^2+(2-1)^2} \Rightarrow$
   $UN=\sqrt{(-11)^2+(1)^2}=\sqrt{121+1}=\sqrt{122}$. Checking the squares of the sides we see
   $UN^2+FN^2=\left(\sqrt{122}\right)^2+\left(\sqrt{32}\right)^2=122+32=154$
   and $UF^2=\left(\sqrt{234}\right)^2=234$. This means $UN^2+FN^2<UF^2$, so this forms an obtuse triangle.

31.
   a. $\left(\frac{4+x}{2},\frac{-2+y}{2}\right)=(6,-13)\Rightarrow \frac{4+x}{2}=6$ and $\frac{-2+y}{2}=-13$. Solving these two equations gives us
   $\frac{4+x}{2}=6 \Rightarrow 4+x=12 \Rightarrow x=8$ and
   $\frac{-2+y}{2}=-13 \Rightarrow -2+y=-26 \Rightarrow y=-24$. So the missing point is $B(8,-24)$.

33.
   b. The radius is the distance from the center to one of the points on the circle:
   $r=\sqrt{(11-4)^2+(7-6)^2} \Rightarrow$
   $r=\sqrt{(7)^2+(1)^2}=\sqrt{49+1}=\sqrt{50}$.

38. Since the width is $3l+2$, when the length is l, we could graph this with one vertex at the origin and two other vertices along the axes.

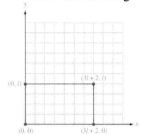

42.
   a. The diagonals are $\overline{AC}$ and $\overline{BD}$. So
   $AC=\sqrt{(-3-2)^2+(2-0)^2} \Rightarrow$
   $AC=\sqrt{(-5)^2+(2)^2} \Rightarrow AC=\sqrt{25+4}=\sqrt{29}$
   and $BD=\sqrt{(2-0)^2+(-3-2)^2} \Rightarrow$
   $BD=\sqrt{(2)^2+(-5)^2} \Rightarrow BD=\sqrt{4+25}=\sqrt{29}$.
   Then $AC=BD \Rightarrow \overline{AC}\cong\overline{BD}$, meaning the diagonals are congruent.
   b. The slope of $\overline{BC}$ is $\frac{2-0}{0-2}=-1$ and the slope of $\overline{AD}$ is $\frac{-3-2}{2-(-3)}=-1$. The bases have the same slopes, so the bases are parallel.
   c. Because they are either horizontal or vertical, the lengths of the legs are quick to find. $AB=3$ and $CD=3$, so the legs of the trapezoid are congruent.

46.
   a. Since this is a rhombus, we know that $\overline{AD}\parallel\overline{BC}$. Since we would move k units left to move from D to A, then we would move k units left to move from C to B. This makes the coordinates $B(p-k,q)$.

b. Since this is a rhombus, we know $AD = DC$.
$AD = \sqrt{(k-0)^2 + (0-0)^2} \Rightarrow$
$AD = \sqrt{k^2 + 0^2} = \sqrt{k^2} = k$ and
$DC = \sqrt{(p-k)^2 + (q-0)^2} \Rightarrow$
$DC = \sqrt{p^2 - 2pk + k^2 + q^2}$. Then
$AD^2 = DC^2 \Rightarrow k^2 = \left(\sqrt{p^2 - 2pk + k^2 + q^2}\right)^2 \Rightarrow$
$k^2 = p^2 - 2pk + k^2 + q^2 \Rightarrow$
$0 = p^2 - 2pk + q^2 \Rightarrow 2pk = p^2 + q^2$.

c. The area of a rhombus is the same as a parallelogram, $A = bh$. In this situation, the base is k and the height is q, so the area is $A = kq$.

48.

a. The slope of $\overline{AC}$ is $\frac{3-1}{2-(-2)} = \frac{1}{2}$ and the slope of $\overline{BD}$ is $\frac{-4-4}{3-(-1)} = -2$. The product of the slopes of the diagonals is -1, so the diagonals of the kite are perpendicular.

b. The midpoint of $\overline{AC}$ is $(0,2)$. The midpoint of $\overline{BD}$ is $(1,0)$. The midpoints of the diagonals are different, so the diagonals do not bisect each other. We can create an equation to represent the points on diagonal $\overline{BD}$. The line has slope -2 and passes through $(1,0)$. The y-intercept would satisfy the equation $0 = -2 \cdot 1 + b \Rightarrow b = 2$, making the equation $y = -2x + 2$ for $-1 \leq x \leq 3$. The midpoint of $\overline{AC}$ is $(0,2)$, and it belongs to diagonal $\overline{BD}$ because $2 = -2 \cdot 0 + 2$. So $\overline{BD}$ bisects $\overline{AC}$.

Section 13.3

3.

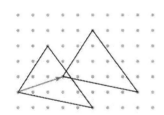

5.

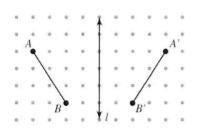

8. The mirror line is the perpendicular bisector of $\overline{PP'}$.

10.

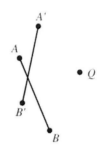

12. This folding confirms that the opposite sides of a rectangle are congruent (and also that the angles are all congruent).

13. The design has point symmetry with a center of rotation at the center of the circle. It has 8 lines of symmetry and an angle of symmetry of $45°$.

17.

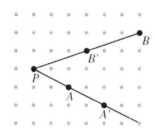

19.

a.

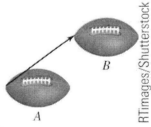

a. If we measure the length of the football on the page and the length of the translation vector, we can create a proportion to find how far the football was moved. This turns out to be about 5.6 cm.

20.
a. Because the image of the line segment is smaller than the original, the scale factor must be less than 1.
b. Measuring the line segments, we can see that $AB \approx 24$ mm and $A'B' \approx 16$ mm, so $k = \dfrac{A'B'}{AB} \approx \dfrac{16}{24} \approx \dfrac{2}{3} \approx 0.67$.

23.

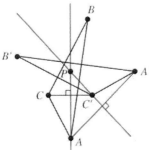

We can find the center of rotation by locating the perpendicular bisectors of $\overline{AA'}$ and $\overline{BB'}$. The center of rotation is P in the diagram.

25.
a. $4 = \dfrac{QA'}{QA} \Rightarrow 4 = \dfrac{QA'}{5} \Rightarrow QA' = 20$ ft

26.
c. $5 = \dfrac{A'B'}{AB} \Rightarrow 5 = \dfrac{A'B'}{7} \Rightarrow A'B' = 35$ cm.

29.
c. We could conjecture that the angle of rotation seems to be $2n$ degrees.

30.
a. Connecting corresponding points in the image and original with a line, and repeating, the center of the size transformation will be where the lines intersect (point P in the figure).

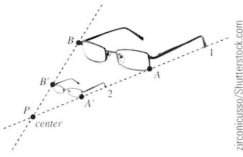

b. Since the image is smaller, the scale factor is less than 1.

c. The scale factor $k = \dfrac{PA'}{PA}$. Use your ruler to measure PA and PA', then divide and you should get close to $k \approx 0.4$.

32. An isometry preserves distances, so $AB = A'B'$, $BC = B'C'$, and $AC = A'C'$. Then $\triangle ABC \cong \triangle A'B'C'$ by the SSS congruence axiom.

35. $AB = \sqrt{(-3-2)^2 + (1-5)^2} \Rightarrow$
$AB = \sqrt{(-5)^2 + (-4)^2} = \sqrt{25+16} = \sqrt{41}$ and
$A'B' = \sqrt{(6-4)^2 + (2-7)^2} \Rightarrow$
$A'B' = \sqrt{(2)^2 + (-5)^2} = \sqrt{4+25} = \sqrt{29}$. This could not have been an isometry since $AB \neq A'B'$, but isometries preserve distance.

41. Answers vary. First, translate the triangle with a translation vector that moves from A to X.

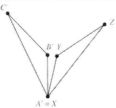

Second, rotate the triangle clockwise an angle of $m\angle B'XY$ about the center A'.

Third, reflect the triangle about the mirror line $\overleftrightarrow{A''B''} = \overleftrightarrow{XY}$.

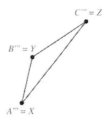

46.

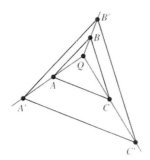

52.

Statement	Reason
1. $PA = PA'$	1. Rotations preserve distance
2. $PB = PB'$	2. Rotations preserve distance
3. $PB = PA + AB$	3. A is between P and B, and the points are collinear.
4. $PB' = PA' + A'B'$	4. A' is between P and B' and the points are collinear.
5. $PB = PA' + A'B'$	5. Substitution from step 2.
6. $PA + AB = PA' + A'B'$	6. Substitution from step 3.
7. $AB = A'B'$	7. Subtraction property of equality using step 1.

58. To find the scale factor, we need to find the lengths of the two segments.

$QB' = \sqrt{(7-1)^2 + (-19-(-1))^2} \Rightarrow$

$QB' = \sqrt{(6)^2 + (-18)^2} = \sqrt{36 + 324} \Rightarrow$

$QB' = \sqrt{360} = 6\sqrt{10}$ and

$QB = \sqrt{(3-1)^2 + (-7-(-1))^2} \Rightarrow$

$QB = \sqrt{(2)^2 + (-6)^2} = \sqrt{4+36} \Rightarrow$

$QB = \sqrt{40} = 2\sqrt{10}$. Then $k = \dfrac{QB'}{QB} = \dfrac{6\sqrt{10}}{2\sqrt{10}} = 3$.

63.

a. The scale factor is $k = \dfrac{B'C'}{BC} = \dfrac{7.5}{6.25} = 1.2$.

Chapter 13 – Review

3.

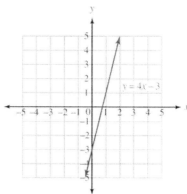

4. l_1 passes through $(1,0)$ and $(-1,-1)$, so the slope is $m = \dfrac{-1-0}{-1-1} = \dfrac{-1}{-2} = \dfrac{1}{2}$. We can find the y-intercept using $0 = \dfrac{1}{2} \cdot 1 + b \Rightarrow b = -\dfrac{1}{2}$. So l_1 is $y = \dfrac{1}{2}x - \dfrac{1}{2}$. l_2 goes through $(-4,0)$ and $(0,-3)$, so the slope is $m = \dfrac{-3-0}{0-(-4)} = \dfrac{-3}{4} = -\dfrac{3}{4}$. We know the y-intercept is $(0, -3)$, so l_2 is $y = -\dfrac{3}{4}x - 3$. l_3 is horizontal and passes through $(0, 6)$, so l_3 is $y = 6$. l_4 is vertical and passes through $(6, 0)$, so l_4 is $x = 6$.

8.

a. With a slope of 3 and y-intercept of $(0,2)$, an equation would be $y = 3x + 2$.

b. With a slope of 2 and x-intercept of $(-3,0)$, we could find the y-intercept using $0 = 2 \cdot (-3) + b \Rightarrow b = 6$. So an equation would be $y = 2x + 6$.

14. The slope of this line is found by putting the line into slope-intercept form: $-3y = 5x - 7 \Rightarrow y = -\dfrac{5}{3}x + \dfrac{7}{3}$.

a. Using the slope, we can see that if the x value decreases by 2 units: $-\dfrac{5}{3} = \dfrac{y}{-2} \Rightarrow 3y = 10 \Rightarrow y = \dfrac{10}{3}$. So the y value would increase by $\dfrac{10}{3}$ units.

15.

a. slope $= \dfrac{\text{change in } y}{\text{change in } x}$, so $-2 = \dfrac{\text{change in } y}{5}$. Then change in $y = -2 \cdot 5 = -10$. Therefore, if the price increases by \$5, then the number of tires sold would decrease by 10.

b. slope $= \dfrac{\text{change in } y}{\text{change in } x}$, so $-2 = \dfrac{\text{change in } y}{-3}$. Then the change in $y = -2 \cdot -3 = 6$. Therefore, if the price decreases by \$3, then the number of tires sold would increase 6.

17.
  a. Neither, since the slopes are $m_1 = 3$ and $m_2 = 18$.
  b. Parallel, since the slopes are $m_1 = -2$ and $m_2 = -2$. The slope of the second line is found with slope-intercept form:
  $10y + 20x = 6 \Rightarrow 10y = -20x + 6 \Rightarrow y = -2x + 0.6$.

20.
  a. A line perpendicular to l would have slope of $-\frac{5}{6}$ so that the product of the two would be -1.
  b. A line parallel to l would have slope of $\frac{6}{5}$, since parallel (non-vertical) lines have the same slope.

26.
  a. $RP = \sqrt{(10-(-2))^2 + (-7-5)^2} \Rightarrow$
  $RP = \sqrt{12^2 + (-12)^2} = \sqrt{144+144} = \sqrt{288}$.
  $RQ = \sqrt{(-2-1)^2 + (5-3)^2} \Rightarrow$
  $RQ = \sqrt{(-3)^2 + (2)^2} = \sqrt{9+4} = \sqrt{13}$.
  $PQ = \sqrt{(1-10)^2 + (3-(-7))^2} \Rightarrow$
  $PQ = \sqrt{(-9)^2 + (10)^2} = \sqrt{81+100} = \sqrt{181}$. The longest side is $\overline{RP}$, so $\angle Q$ is the largest angle.
  b. $RP = \sqrt{(1-2)^2 + (-2-8)^2} \Rightarrow$
  $RP = \sqrt{(-1)^2 + (-10)^2} = \sqrt{1+100} = \sqrt{101}$.
  $RQ = \sqrt{(3-2)^2 + (5-8)^2} \Rightarrow$
  $RQ = \sqrt{(1)^2 + (-3)^2} = \sqrt{1+9} = \sqrt{10}$.
  $PQ = \sqrt{(3-1)^2 + (5-(-2))^2} \Rightarrow$
  $PQ = \sqrt{(2)^2 + (7)^2} = \sqrt{4+49} = \sqrt{53}$. The longest side is $\overline{RP}$, so $\angle Q$ is the largest angle.

28.
  a. $AB = \sqrt{(-3-(-1))^2 + (2-4)^2} \Rightarrow$
  $AB = \sqrt{(-2)^2 + (-2)^2} = \sqrt{4+4} = \sqrt{8}$.
  $AC = \sqrt{(-3-1)^2 + (2-(-2))^2} \Rightarrow$
  $AC = \sqrt{(-4)^2 + (4)^2} = \sqrt{16+16} = \sqrt{32}$.
  $BC = \sqrt{(-1-1)^2 + (4-(-2))^2} \Rightarrow$
  $BC = \sqrt{(-2)^2 + (6)^2} = \sqrt{4+36} = \sqrt{40}$. Now we can check $(AB)^2 + (AC)^2$ and $(BC)^2$ to see if this is a right triangle. $BC^2 = 40$ and $(AB)^2 + (AC)^2 = 8 + 32 = 40$, so $(AB)^2 + (AC)^2 = (BC)^2$. This is a right triangle by the converse of the Pythagorean theorem.

31. First, find the midpoint of $\overline{AB}$: $\left(\frac{-3+1}{2}, \frac{2+4}{2}\right) = (-1, 3)$. The slope of $\overline{AB}$ is $m = \frac{2-4}{-3-1} = \frac{-2}{-4} = \frac{1}{2}$, so the slope of the perpendicular line will be -2. We can find the y-intercept using the point $(-1, 3)$ and $3 = -2 \cdot (-1) + b \Rightarrow b = 3 - 2 = 1$. An equation of the line is $y = -2x + 1$.

34. Looking at the picture, we see the center is $(-2, 3)$ and another point on the circle is $(2, 5)$. We can find the radius of the circle by finding the distance between these two points:
$r = \sqrt{(-2-2)^2 + (3-5)^2} \Rightarrow$
$r = \sqrt{(-4)^2 + (-2)^2} = \sqrt{16+4} = \sqrt{20}$. Writing this in standard form, we have
$(x-(-2))^2 + (y-3)^2 = \left(\sqrt{20}\right)^2$

36. Since this is a parallelogram, side $\overline{AB}$ will be parallel to and the same length as $\overline{CD}$. To move from B to A, we travel 1 unit right and 7 units up. So to find point D, we need to start at point C and travel 1 unit right and 7 units up which ends at $D(-1, 6)$.

38. Since the width is $5l-2$ when the length is l, we could place one vertex at the origin and two sides along the axes to create a simpler picture.

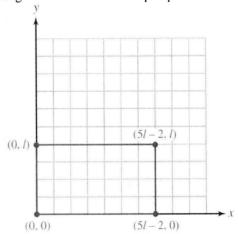

118

40. Answers vary. To draw a parallelogram with area of 10 square units, we could have a base of 5 units and a height of 2 units.

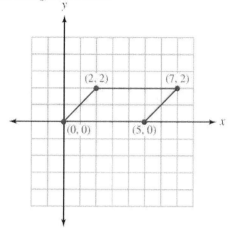

42.
a. The diagonals are $\overline{AC}$ and $\overline{BD}$. So
$AC = \sqrt{(-3-2)^2 + (-1-2)^2} \Rightarrow$
$AC = \sqrt{(-5)^2 + (-3)^2} = \sqrt{25+9} = \sqrt{34}$ and
$BD = \sqrt{(-3-2)^2 + (2-(-1))^2} \Rightarrow$
$BD = \sqrt{(-5)^2 + (3)^2} = \sqrt{25+9} = \sqrt{34}$. Then $AC = BD$, so $\overline{AC} \cong \overline{BD}$, meaning the diagonals are congruent.

b. The midpoint of $\overline{AC}$ is $\left(\frac{-3+2}{2}, \frac{-1+2}{2}\right) = \left(\frac{-1}{2}, \frac{1}{2}\right)$ and the midpoint of $\overline{BD}$ is $\left(\frac{-3+2}{2}, \frac{2-1}{2}\right) = \left(\frac{-1}{2}, \frac{1}{2}\right)$. Since the diagonals have the same midpoint, the diagonals bisect each other.

48. This translation moves 2 units left and 9 units down (using A and A' as a guide), so the image of B will be 2 units left and 9 units down, ending at $B'(4, -5)$.

53. Since the images are mirror images, this must have been a reflection.

55.
a. To find the center of rotation, we need to find a point that is the same distance from A and A', so we need to find the perpendicular bisector of $\overline{AA'}$. Repeat the process with B and B'. Where these two perpendicular bisectors intersect is the center of rotation, call it P.

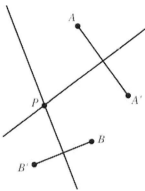

b. Using a protractor, the angle is about $60°$.

57.
a. There are 5 shapes, so the angle of symmetry is $\frac{360}{5} = 72°$.
b. There are 6 shapes, so the angle of symmetry is $\frac{360}{6} = 60°$.

58.
a. one line of symmetry   b. two lines of symmetry

60.
a.

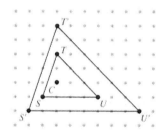

b. $S'T' = \sqrt{(2)^2 + (6)^2} = \sqrt{4+36} = \sqrt{40}$.
$ST = \sqrt{(1)^2 + (3)^2} = \sqrt{1+9} = \sqrt{10}$.
$\frac{S'T'}{ST} = \frac{\sqrt{40}}{\sqrt{10}} = \frac{2\sqrt{10}}{\sqrt{10}} = 2$.

64.
a. $QA' = k \cdot QA \Rightarrow QA' = 4 \cdot 5 = 20$ cm.

65. The scale factor can be found using $k = \frac{QA'}{QA}$, so we can find the two lengths and then divide.
$QA' = \sqrt{(14-3)^2 + (-49-(-5))^2} \Rightarrow$
$QA' = \sqrt{(11)^2 + (44)^2} = \sqrt{11^2 + 4^2 \cdot 11^2} \Rightarrow$
$QA' = \sqrt{17 \cdot 11^2} = 11\sqrt{17}$.
$QA = \sqrt{(5-3)^2 + (-13-(-5))^2} \Rightarrow$

$QA = \sqrt{(2)^2 + (-8)^2} = \sqrt{4+64} \Rightarrow$
$QA = \sqrt{68} = 2\sqrt{17}$. Then the scale factor is
$k = \dfrac{QA'}{QA} = \dfrac{11\sqrt{17}}{2\sqrt{17}} = \dfrac{11}{2} = 5.5$.

67. Pick two points on one owl, and then the corresponding points on the other owl. Measure the length of the segment in one owl, and then repeat for the other. Divide these values to obtain the result $k \approx 1.6$.

Chapter 13 Test

2.
   a. With a slope of -12, the line will look like $y = -12x + b$. Since it must also pass through $(4, -10)$, we have
   $-10 = -12 \cdot 4 + b \Rightarrow b = -10 + 48 = 38$. The equation would be $y = -12x + 38$.

5.
   a. The slope is 6.50, which represents the rate of change. $6.50 = \dfrac{94.25}{x} \Rightarrow 6.50x = 94.25 \Rightarrow x = 14.5$. The number of feet of fence painted would increase by 14.5 ft.

6.
   a. Parallel. Find the slope of the second line and we see $8y = 56x + 32 \Rightarrow y = 7x + 4$. So the slope of both lines is 7, making the lines parallel.

10. $RP = \sqrt{(8-(-3))^2 + (1-3)^2} \Rightarrow$
    $RP = \sqrt{(11)^2 + (-2)^2} = \sqrt{121+4} = \sqrt{125}$.
    $RQ = \sqrt{(3-8)^2 + (-1-1)^2} \Rightarrow$
    $RQ = \sqrt{(-5)^2 + (-2)^2} = \sqrt{25+4} = \sqrt{29}$.
    $PQ = \sqrt{(-3-3)^2 + (3-(-1))^2} \Rightarrow$
    $PQ = \sqrt{(-6)^2 + (4)^2} = \sqrt{36+16} = \sqrt{52}$. Now we can check $(RQ)^2 + (PQ)^2$ and $(RP)^2$ to determine the type of triangle. $(RP)^2 = 125$ and $(RQ)^2 + (PQ)^2 = 29 + 52 = 81$, so $(RQ)^2 + (PQ)^2 < (RP)^2$. $\triangle PQR$ is an obtuse triangle.

13.
   a. From the picture, we can see that $AD = k$. Since the shape is a rhombus, we know that the sides will all be congruent and parallel. Since D is k units right of A, then C will be k units right of B, with coordinates of $C(p+k, q)$.

14. The midpoint of $\overline{AC}$ is $\left(\dfrac{1+0}{2}, \dfrac{4-6}{2}\right) = \left(\dfrac{1}{2}, -1\right)$ and the midpoint of $\overline{BD}$ is $\left(\dfrac{6-5}{2}, \dfrac{-4+2}{2}\right) = \left(\dfrac{1}{2}, -1\right)$. Since the diagonals have the same midpoint, the diagonals bisect each other.

16.
   a. In order, A is mapped to W with a reflection, and then to H with a translation.

20.
   a. There are 7 wedge shapes, so there will be 7 lines of symmetry.
   b. The angle of symmetry is $\dfrac{360}{7} = 51\tfrac{3}{7} \approx 51.43$.

22. The scale factor can be found using $k = \dfrac{QA'}{QA}$, so we can find the two lengths and then divide.
    $QA' = \sqrt{(4-(-4))^2 + (10-(-2))^2} \Rightarrow$
    $QA' = \sqrt{(8)^2 + (12)^2} = \sqrt{64+144} = \sqrt{208}$.
    $QA = \sqrt{(-2-(-4))^2 + (1-(-2))^2} \Rightarrow$
    $QA = \sqrt{(2)^2 + (3)^2} = \sqrt{4+9} = \sqrt{13}$. Then the scale factor is $k = \dfrac{QA'}{QA} = \dfrac{\sqrt{208}}{\sqrt{13}} = \dfrac{4\sqrt{13}}{\sqrt{13}} = 4$.

24.
   a. The scale factor can be found using $k = \dfrac{A'B'}{AB} = \dfrac{10.32}{4.30} = 2.4$.

CPSIA information can be obtained
at www.ICGtesting.com
Printed in the USA
FFOW02n1837310714
6630FF